A Wheel Through the Primes

A Concrete Tour of Analytic Number Theory

Book One of The Primes Trilogy

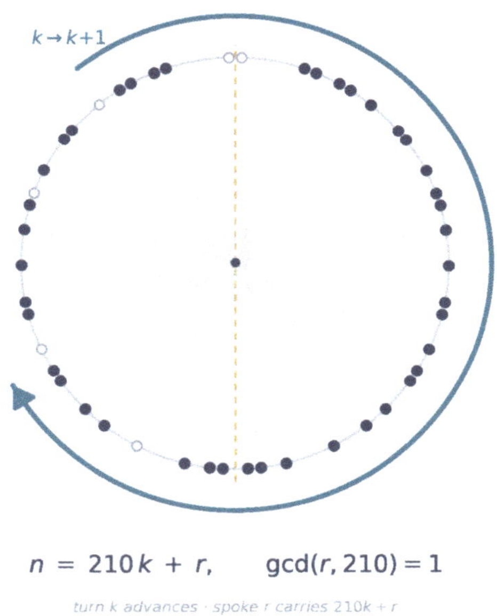

John Maggio

Efficient Intelligence, LLC

2026

Published by Efficient Intelligence, LLC

Paperback ISBN: 979-8-9967325-0-0
Hardcover ISBN: 979-8-9967325-1-7
Ebook ISBN: 979-8-9967325-2-4

First edition, 2026   Printed in the United States of America

Advance in Peace

Preface

This is a book about the prime numbers, told through a single object: a wheel of forty-eight spokes, the residues coprime to $210 = 2 \cdot 3 \cdot 5 \cdot 7$. Everything in it — the way primes spread across the spokes, the faint bias in their race, the mirror that pairs each spoke with another, the diffraction grating that sifts them, the zeros of the zeta function reappearing as their spectrum, the gas of prime "modes," the climb up the primorials, the twin slots — is standard analytic number theory. The wheel is not a new theory. It is a place to stand where the standard theory can be seen at human scale.

Two things are worth saying at the outset. The book promises that the machinery is real and that you will meet it concretely, not in outline: Dirichlet characters as the harmonics of the wheel, the explicit formula as genuine wave interference, the Hardy–Littlewood singular series as fifteen surviving slots out of 210. And it promises the opposite of a triumphal ending. The tour does not solve the Riemann Hypothesis, prove the twin-prime conjecture, or make factoring easy. It ends, deliberately, at those three walls — because that is where the subject actually ends, and because seeing exactly where a wall stands, and why, is worth more than a story that pretends to pass it.

The wheel earns its place by being small. 210 is the smallest modulus that resolves the first four primes at once, so it is the smallest wheel on which the whole early sieve is visible in a single turn. That legibility is the entire source of its specialness. The structures it reveals belong to the integers, not to 210, and the book closes by saying so plainly.

How to read this

The chapters build, but loosely. Chapter 1 sets up the wheel and is assumed everywhere after it; beyond that, each part can be sampled on its own. A reader who wants the spectral heart of the subject can go from Chapter 1 straight to Part III; a reader who wants the elementary combinatorics can stay in Parts I and IV. The figures are meant to be read, not skipped — much of the argument lives in them. Each figure is accompanied in the text by the short Python program that produced it, printed in full so it can be rerun rather than trusted; this book ships no separate code files. A glossary at the back collects the recurring vocabulary.

Acknowledgements

The author gratefully acknowledges the use of Anthropic's Claude as a computational and drafting assistant during the preparation of this work. Any mathematical results or interpretations presented here remain entirely the responsibility of the author. — J.M.

Contents

Part I

The object

Chapter 1

The Mod-210 Wheel

This is the opening chapter of a book that studies the prime numbers through a single concrete object: the wheel modulo $210 = 2 \cdot 3 \cdot 5 \cdot 7$*. The book claims no new result; its aim is unity and clarity — to gather a scattered body of classical number theory onto one stage, so that each idea can be seen as a feature of the same picture. Here we build the stage. The integers coprime to* 210 *form a group of* $\varphi(210) = 48$ *residue classes — the "spokes" of the wheel — and every prime larger than* 7 *lands on one of them. We record the structure of this group,* $(\mathbb{Z}/210\mathbb{Z})^\times \cong \mathbb{Z}/2 \times \mathbb{Z}/4 \times \mathbb{Z}/6$*; the wheel as a sieve, with the surviving density* $\varphi(M)/M = \prod_{p|M}(1 - 1/p)$*; and the reflection* $r \mapsto 210 - r$ *that returns in every later chapter.*

The later chapters then act on this stage. Dirichlet's theorem spreads the primes evenly across the 48 *spokes, while a closer look exposes the persistent Chebyshev tilt — a prime race among the classes. The wheel's seven real characters arrange into a conductor lattice split by parity, and the reflection* $r \mapsto 210 - r$ *is unmasked as the elementary shadow of the functional equation* $s \mapsto 1 - s$*. The primes are then read off the zeros of the zeta function as a diffraction pattern; the zeros themselves repel like the eigenvalues of a random matrix; and the squarefree integers assemble into a supersymmetric "primon gas" whose bookkeeping is the Möbius function. Climbing the primorials yields a conservation law and Mertens' theorem; the repeating decimals of the wheel — a structure already in Gauss — split the spokes into two cyclic families; and the book closes on the twin primes, where the same wheel that organizes everything so cleanly runs headlong into the parity barrier that no elementary method has crossed.*

Everything is elementary and classical. The wheel is a lens, not a law: it renders a great deal of known mathematics visible at once, and it is candid about where the difficulty actually lives — not in the structure of the wheel, which is settled, but in how the primes distribute across it, where the oldest questions remain open. This first note fixes the stage cleanly, before the others act on it.

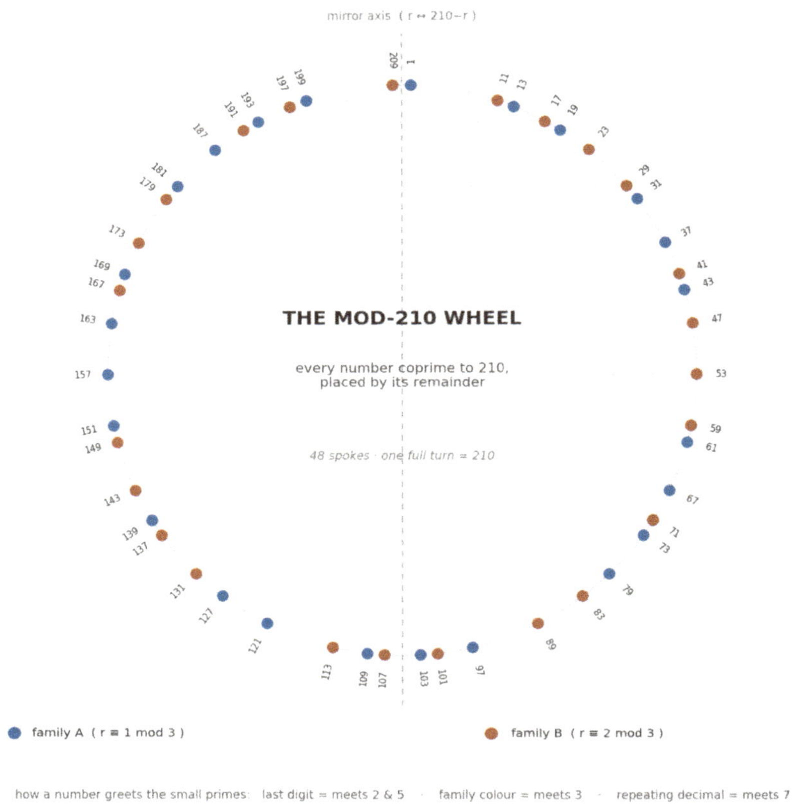

Figure 1.1: The whole subject on a single face. Every number coprime to 210 is placed by its remainder; color marks the family ($r \equiv 1$ versus $r \equiv 2 \bmod 3$), and the vertical line is the mirror $r \mapsto 210 - r$, across which the two families swap. The line at the foot decodes the rest: a number's last digit records how it meets 2 and 5, its color how it meets 3, and its repeating decimal how it meets 7. Everything in this chapter and the ones that follow is a closer look at one feature of this one picture.

1.1 Scope

This chapter is foundational and entirely classical [1, 2]. It introduces no result; it fixes notation and the basic structure of the mod-210 wheel so that the chapters that follow — on the prime race, the real characters and their conductor lattice, the zeros as a diffraction pattern, and the supersymmetric primon gas — can each act on a stage that has already been described.

1.2 The modulus and its units

The modulus is

$$210 = 2 \cdot 3 \cdot 5 \cdot 7, \tag{1.1}$$

the product of the first four primes — the fourth *primorial*. The residues modulo 210 that are coprime to it form the unit group $(\mathbb{Z}/210\mathbb{Z})^\times$, whose size is Euler's totient

$$\varphi(210) = \varphi(2)\,\varphi(3)\,\varphi(5)\,\varphi(7) = 1 \cdot 2 \cdot 4 \cdot 6 = 48. \tag{1.2}$$

These 48 classes are the spokes of the wheel (Fig. 1.2). By the Chinese Remainder Theorem the group factors as

$$(\mathbb{Z}/210\mathbb{Z})^\times \;\cong\; (\mathbb{Z}/2)^\times \times (\mathbb{Z}/3)^\times \times (\mathbb{Z}/5)^\times \times (\mathbb{Z}/7)^\times \;\cong\; \mathbb{Z}/2 \times \mathbb{Z}/4 \times \mathbb{Z}/6, \tag{1.3}$$

of order $2 \cdot 4 \cdot 6 = 48$. It is not cyclic — 210 has no primitive root — which is why the wheel carries several independent quadratic characters rather than one; that multiplicity is the subject of the conductor-lattice note.

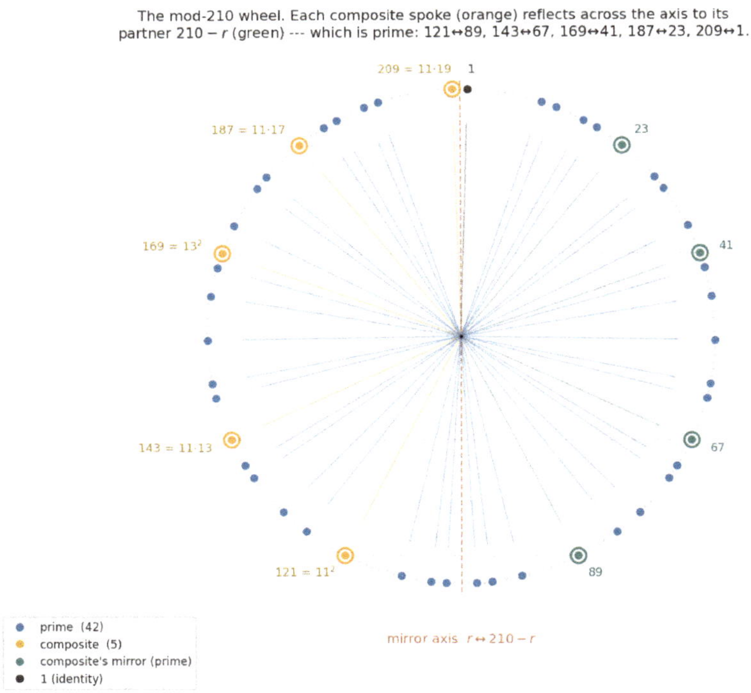

Figure 1.2: The mod-210 wheel. The 48 residues coprime to 210 are the spokes; the remaining positions (multiples of 2, 3, 5, or 7) are the faint dots on the rim. The spokes are colored by type: of the 48, one is the identity, 42 are primes, and five are composite (orange)—the products of two primes ≥ 11 that remain below 210, namely $121 = 11^2$, $143 = 11 \cdot 13$, $169 = 13^2$, $187 = 11 \cdot 17$, and $209 = 11 \cdot 19$. Every prime greater than 7 is a spoke, but so is each of these composites. The dashed axis is the reflection $r \mapsto 210 - r$ of Sec. 1.4. Reflecting each composite across it gives its partner $210 - r$ (green, with faint chords): $121 \leftrightarrow 89$, $143 \leftrightarrow 67$, $169 \leftrightarrow 41$, $187 \leftrightarrow 23$, and $209 \leftrightarrow 1$. Every one of those partners is prime (or the identity) — the reflection keeps a residue coprime to 210 but does not keep it composite.

1.3 The wheel as a sieve

The reason the primes live here is immediate: a prime $p > 7$ is divisible by none of $2, 3, 5, 7$, so $\gcd(p, 210) = 1$ and p occupies one of the 48 unit classes. The first four primes $2, 3, 5, 7$ are exactly the teeth of the sieve — they are the primes *removed* to build the wheel — so apart from those four, the entire prime sequence is distributed among the 48 spokes. This is the elementary content of *wheel factorization*: to hunt for primes one need only examine the residues coprime to 210, a little under a quarter of all integers.

The spokes are not all prime, and it would misrepresent the wheel to draw them as if they were. Of the 48, one is the identity 1, forty-two are the primes between 11 and 199, and five are composite: the products $121 = 11^2$, $143 = 11 \cdot 13$, $169 = 13^2$,

$187 = 11 \cdot 17$, $209 = 11 \cdot 19$ of two primes at least 11 that stay below 210 (orange in Fig. 1.2). The wheel selects every residue coprime to 210 whatever its factorization; in this range the primes are simply the large majority.

The exact fraction is the density of coprime residues,

$$\frac{\varphi(210)}{210} = \frac{48}{210} = \frac{8}{35} \approx 0.229 = \prod_{p|210}\left(1 - \frac{1}{p}\right) = \left(1-\tfrac{1}{2}\right)\left(1-\tfrac{1}{3}\right)\left(1-\tfrac{1}{5}\right)\left(1-\tfrac{1}{7}\right). \quad (1.4)$$

Adding more primes to the primorial thins the wheel further (Fig. 1.3); by Mertens' theorem the density $\prod_{p\le y}(1-1/p)$ decays like $e^{-\gamma}/\log y$, slowly but without bound. The choice of 210 is a pragmatic sweet spot: large enough that the four smallest primes are swept away and the structure is rich, small enough that all 48 classes fit on a single page.

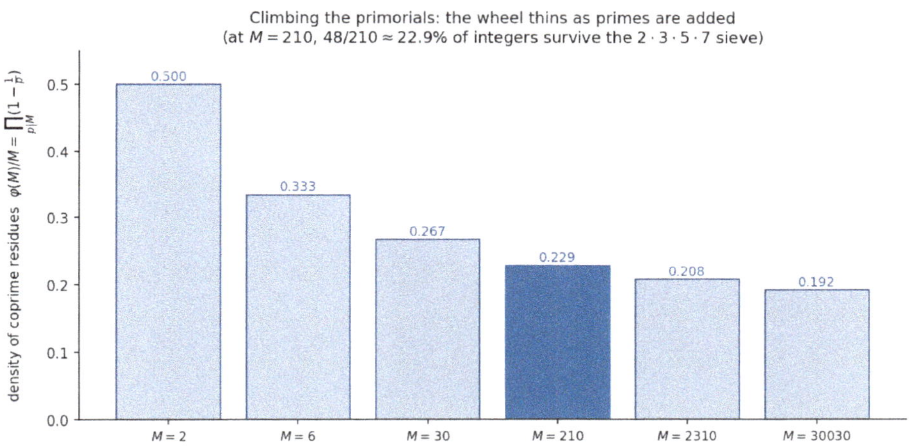

Figure 1.3: The density $\varphi(M)/M = \prod_{p|M}(1-1/p)$ of residues coprime to successive primorials. At $M = 210$ (highlighted) just under 23% of integers survive the $2 \cdot 3 \cdot 5 \cdot 7$ sieve. The decay continues without bound, at the slow Mertens rate.

1.4 The mirror

One symmetry of the wheel deserves to be named at the outset, because it recurs in every later chapter. The reflection

$$r \longmapsto 210 - r \pmod{210} \quad (1.5)$$

is an involution that preserves coprimality, since $\gcd(210-r, 210) = \gcd(r, 210)$. Its only fixed points are $r = 0$ and $r = 105$, and neither is a unit; so the 48 spokes split cleanly into 24 mirror pairs $\{r,\, 210-r\}$. In Fig. 1.2 this is the reflection across the dashed vertical axis.

The reflection keeps a residue coprime to 210, but it need not keep it prime: it sends each of the five composite spokes to a prime, $121 \leftrightarrow 89$, $143 \leftrightarrow 67$, $169 \leftrightarrow 41$,

$187 \leftrightarrow 23$, and $209 \leftrightarrow 1$. The mirror is a fact about residues, not about factorization — a distinction worth keeping in view, since the later notes lean on it.

This single reflection wears several costumes downstream. In the prime-race note it is the axis along which Chebyshev's bias is measured, separating the quadratic residues from the non-residues. In the character note it is the parity $\chi(-1) = \pm 1$, because $\chi(210 - r) = \chi(-r) = \chi(-1)\chi(r)$, sorting the characters into even and odd. And in the deepest reading it is the shadow of the functional equation's reflection $s \mapsto 1 - s$. Seeing it first as a plain geometric fold on the wheel makes its later appearances easier to recognize.

1.5 The wheel turning

The wheel of Fig. 1.2 is a single turn. The integers do not stop at 210: the spoke at residue r carries the number $210k + r$ on turn k, so the same 48 spokes are reused turn after turn, with only the leading multiple of 210 advancing. The later turns are worth a look, because the pattern on the spokes both persists and quietly changes.

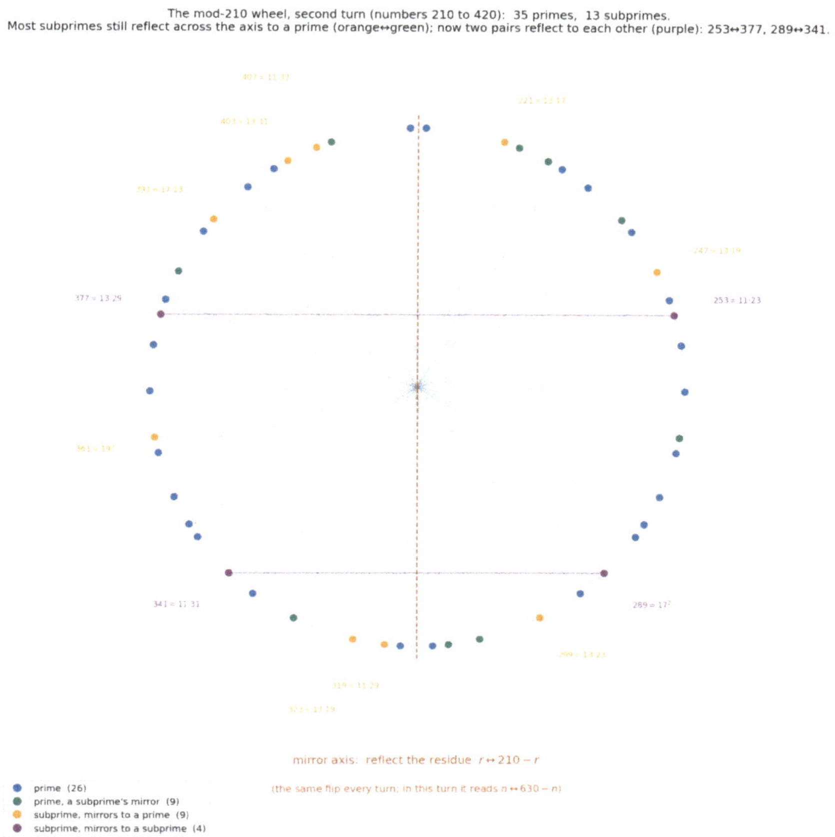

Figure 1.4: The wheel's second turn, the numbers 210 to 420 (the spoke at residue r now carries $210 + r$). The primes have thinned to 35 and the subprimes risen to 13. As on the first turn most subprimes still reflect across the axis onto a prime (orange onto green); the two purple pairs, $253 \leftrightarrow 377$ and $289 \leftrightarrow 341$, are the first composite spokes whose mirror is *also* composite.

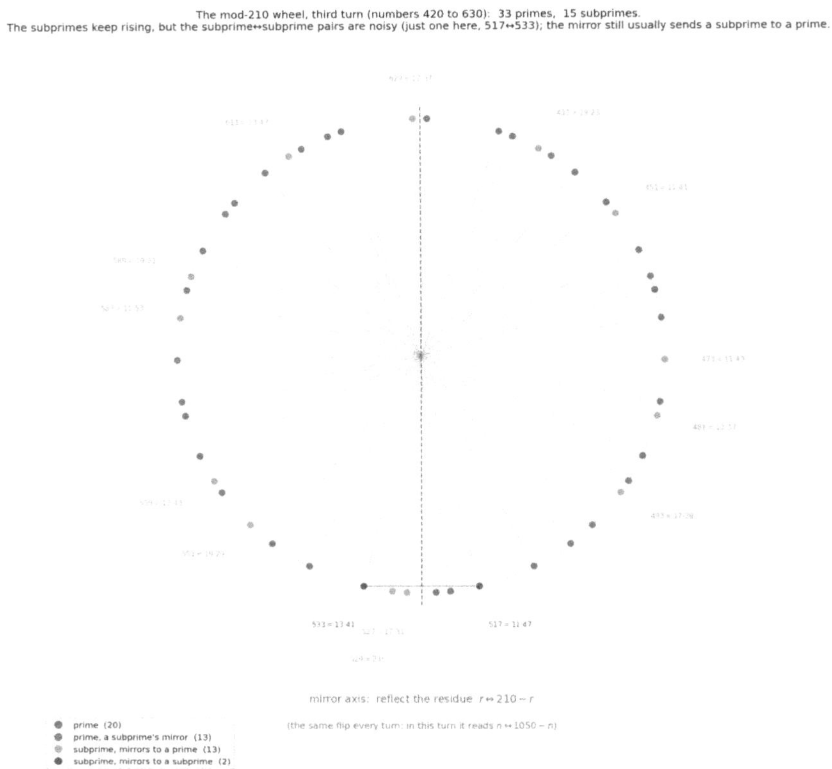

Figure 1.5: The third turn, 420 to 630: 33 primes and 15 subprimes. The thinning continues, while the subprime-onto-subprime pairs are noisy from turn to turn (just one here, $517 \leftrightarrow 533$). The mirror still usually carries a subprime to a prime.

Two things happen as the wheel turns. First, the primes thin: 42 primes on the first turn, 35 on the second, 33 on the third (Figs. 1.4 and 1.5), with the subprimes filling the vacated spokes. This is the prime number theorem seen one turn at a time, the prime density falling like $1/\log$. Second, the mirror of Sec. 1.4 keeps acting, but its character clarifies. On the first turn every composite spoke reflected onto a prime, which can make the reflection look as though it inverts compositeness. From the second turn onward, pairs of composite spokes begin reflecting onto *each other* (Fig. 1.4). The mirror is not inverting primality and never was; it is simply indifferent to it, sometimes sending a composite to a prime and sometimes to another composite. Primality is not among its invariants.

Beneath the thinning lies a strictly periodic mechanism. Fix a spoke r and ask when a prime $p \geq 11$ divides the number sitting on it. Since $p \nmid 210$, we have $210 \equiv c_p \pmod{p}$ for some nonzero c_p, so the residue $210k + r \bmod p$ advances by c_p each turn and reaches 0 once every p turns. Each prime is therefore a metronome with its own period: 11 strikes a given spoke every 11 turns, 13 every 13, 17 every 17. Placing the spokes at their true values straightens these beats into exact diagonals of slope $-c_p$, namely -1, -2, and -6 for 11, 13, and 17 (Fig. 1.6). A spoke is composite on a given turn exactly when at least one metronome strikes it, and prime only when it dodges every one up to $\sqrt{210k + r}$.

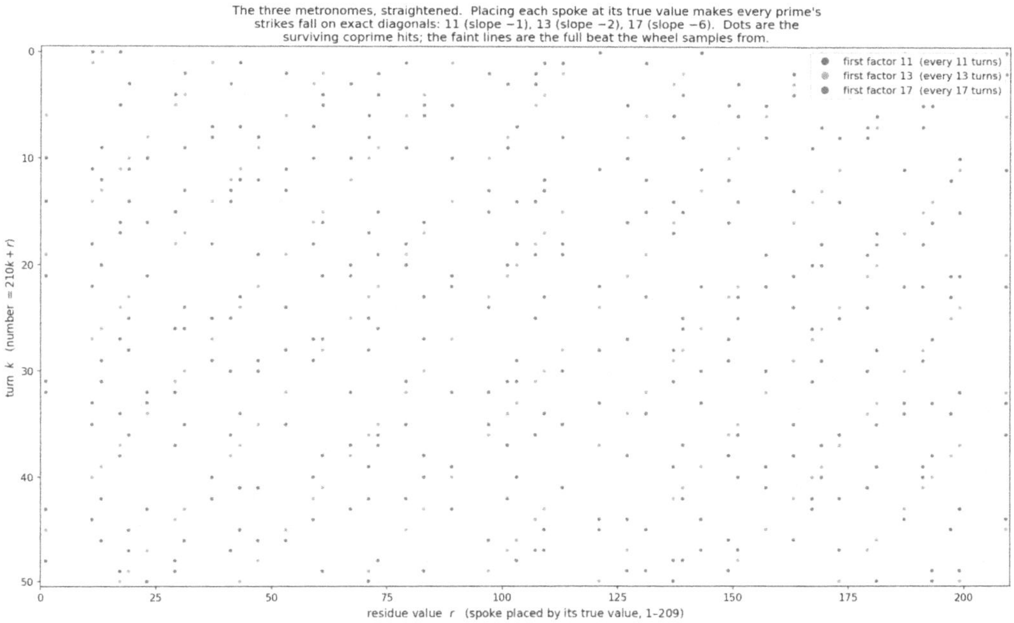

Figure 1.6: The three smallest metronomes across fifty turns. With each spoke placed at its true residue value, the multiples of 11, 13, and 17 fall on exact diagonals of slope $-(210 \bmod p)$. The dots are the coprime hits the wheel keeps; the faint lines are the full periodic beat they are sampled from. The blank space between the stripes is where the primes live.

Pulled apart one metronome to a panel (Fig. 1.7), the same beats become unmistakable. Each prime rides a single family of parallel diagonals, and the larger its step c_p, the faster it sweeps across all 48 spokes: 17, stepping by 6, lays down flatter and more tightly spaced stripes than 11, which steps by 1. Each panel is one prime's entire contribution to the sieve, drawn in isolation.

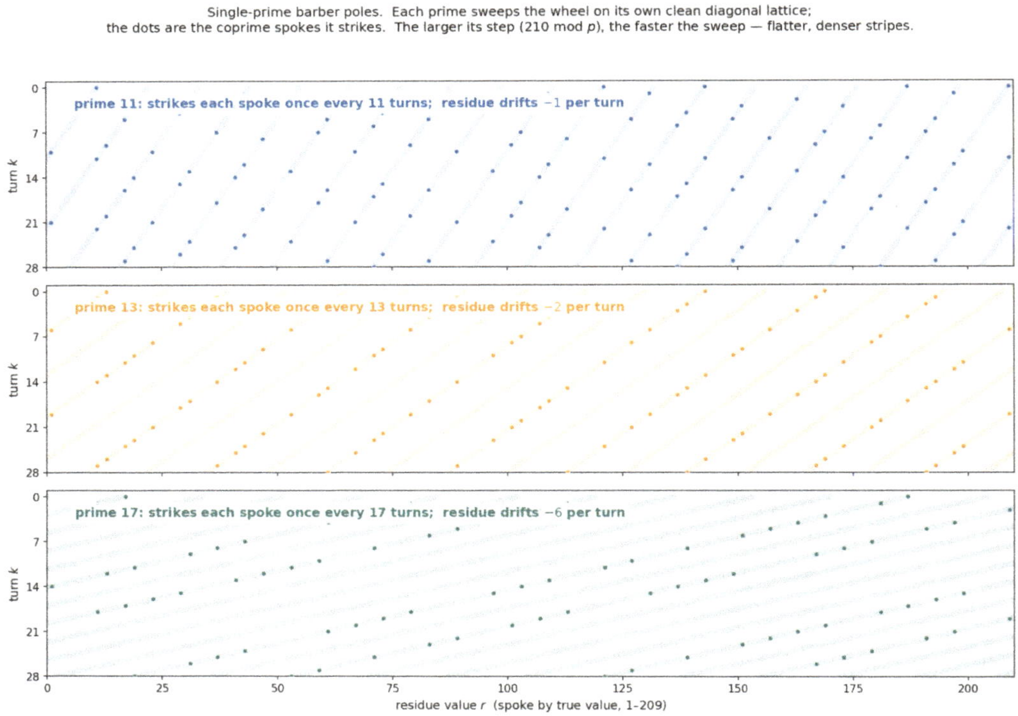

Figure 1.7: The three smallest metronomes separated one to a panel. Each prime p strikes a coprime spoke (the dots) once every p turns, drifting by $-(210 \bmod p)$ spokes per turn, so its hits ride a single family of parallel diagonals — the faint bands are the full periodic beat they are sampled from. The step grows with p, namely -1, -2, and -6 for 11, 13, and 17, so the larger primes sweep the wheel faster: flatter, more crowded stripes.

This is the boundary the wheel keeps drawing, now laid out in time. Any *fixed* set of metronomes is perfectly periodic: 11 and 13 together repeat every $11 \cdot 13 = 143$ turns, and admitting one more prime is precisely the step from one primorial modulus to the next, $210 \to 2310 \to 30030$, which is the subject of the chapter on climbing the primorials. But genuine primality requires every metronome up to $\sqrt{n}$, and that set never stops growing; a new, incommensurate beat keeps joining, so the full pattern never closes into a finite period. The composites are pure clockwork — one could mark every one of them in advance from the metronomes alone. The primes are the spokes no metronome ever claims, and that leftover is the one thing all this periodic machinery cannot itself put a period to. The wheel turns systematically; what it leaves behind does not.

1.6 Two sides of one coin

Every spoke on the wheel, in every turn, carries exactly one number, and that number is either prime or a product of primes that are themselves at least 11; call the latter the *subprimes* of the wheel. The two sets are exact complements: a spoke is prime precisely when it is not a subprime, so the primes are nothing more than the gaps the subprimes leave, and naming either set names the other (Fig. 1.8). This is what a sieve is — Eratosthenes finds the primes by marking every composite and keeping what survives. The only number on the wheel that is neither is the unit 1, sitting alone on the first spoke.

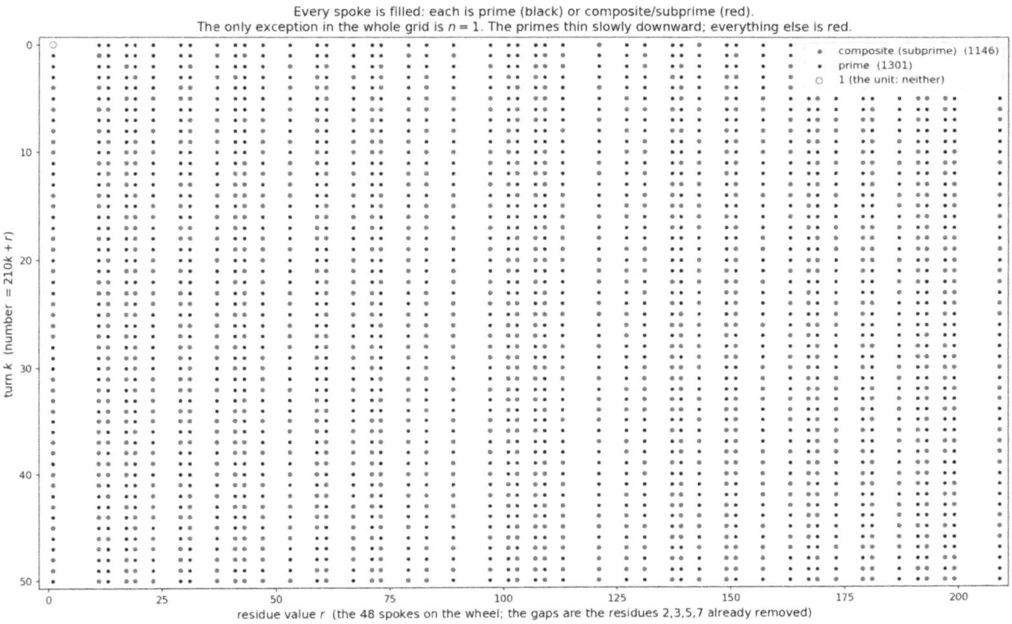

Figure 1.8: Fifty turns of the wheel, every spoke filled: prime (black) or subprime (red). The vertical bands of empty space are not missing spokes; they are the residues divisible by 2, 3, 5, or 7, removed before the wheel begins. Along the 48 true spokes nothing is ever blank, and $1301 + 1146 + 1 = 2448 = 48 \times 51$ accounts for every number. Red gains slowly on black going down: subprimes grow more common as the supply of small prime factors widens, the same $1/\ln n$ thinning of the primes seen from the other side.

The two faces of this coin are not peers, however. The subprimes are *generated*: each is a product of smaller wheel-primes, so the whole red set can be built forwards by multiplication (11^2, $11 \cdot 13$, $13^2, \ldots$), which is exactly the periodic clockwork of the metronomes above. The primes carry no such rule; they are defined only negatively, as the numbers that are *not* such products. The red side is therefore built out of the black side — the primes are the source and the subprimes their shadow. That lopsidedness is the familiar asymmetry of the subject: forming a subprime is a moment's multiplication, while recovering the primes a given subprime was built from is integer factorization, for which no efficient method is known. The duality is perfect; the direction is not.

Stripped of its cheap structure, the leftover looks startlingly like noise, and this too is a precise phenomenon. The primes are *pseudorandom*: fully deterministic, yet passing most statistical tests one can put to them. Cramér made the resemblance quantitative by modeling each integer n as prime independently with probability $1/\ln n$ [3]; on the wheel, where the residues coprime to 210 are already $210/48 = 4.375$ times denser in primes, the matched rate is $(210/48)/\ln n$. Figure 1.9 sets the real wheel beside a single draw of this coin-flip model at that rate. The two are hard to tell apart, which is the point: the model knows nothing about which numbers are actually prime, only their density, yet reproduces their texture. What it cannot reproduce is their rigidity. The real primes are not random — they are forced — and the gap between *behaves like a coin flip* and *is a coin flip* is precisely where the Riemann Hypothesis and the twin-prime conjecture live, and why they remain open.

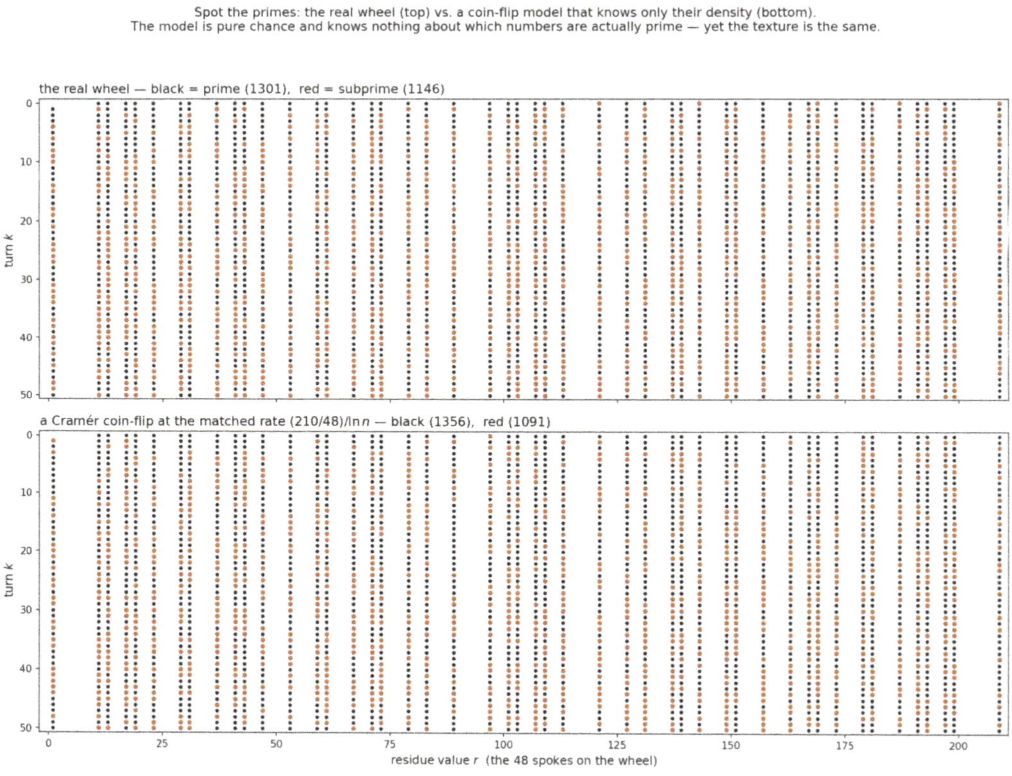

Figure 1.9: The real wheel (top) beside one draw of a Cramér coin-flip (bottom), each number coloured prime with probability $(210/48)/\ln n$ and capped at 1. The model produced 1356 primes against the wheel's 1301 — a match in density, and to the eye a match in texture, though the model is pure chance and the wheel is wholly determined. The visible difference is statistical, not structural; the model's one genuine failure is that it cannot know the primes are not, in fact, being flipped.

This same density reaches further than the two counts. Applied to the wheel's mirror pairs (Sec. 1.4), it fixes how the two reflected residues fall by primality, turn after turn—a small effective model, with the prime number theorem as its only input, set out in the chapter on the mirror-pair populations.

1.7 What this sets up

With the stage fixed, each later chapter puts something in motion on it. The prime race watches the primes distribute themselves across the 48 spokes and measures the slight, persistent bias in that distribution. The conductor-lattice chapter studies the quadratic characters the non-cyclic group carries, and organizes them by their conductors. The diffraction chapter passes to the spectral side, drawing the Riemann and Dirichlet zeros and showing their Fourier transform lands on the primes. The supersymmetric-gas chapter recovers the Möbius function as the partner of those same L-functions. The mirror-pair chapter stays closest to this stage, carrying the Cramér density just above into a count of how the reflection's prime–prime, prime–subprime, and subprime–subprime pairs populate as the primes thin. Every one of them returns, sooner or later, to the 48 spokes and the one mirror described here.

Reproducing the figures

```
from math import gcd
M = 210
units = [r for r in range(M) if gcd(r, M) == 1]
print(len(units))                        # 48
print(units[:6], units[-3:])             # [1, 11, 13, 17, 19, 23] ... [197, 199, 209]

# density of coprime residues for the primorials
prod = 1.0
for p in [2, 3, 5, 7, 11, 13]:
    prod *= (1 - 1/p)
    print(p, round(prod, 4))             # 0.5, 0.3333, 0.2667, 0.2286, 0.2078, 0.1918

# the mirror pairs
pairs = sorted({tuple(sorted((r, (M - r) % M))) for r in units})
print(len(pairs))                        # 24 mirror pairs

# metronome diagonals: the coprime spokes a prime p strikes on turn k
def strikes(p, k):
    step = M % p                         # 11 -> 1, 13 -> 2, 17 -> 6
    return [r for r in units if (r + step * k) % p == 0]
for p in (11, 13, 17):
    print(p, strikes(p, 0))              # 11 [11,121,143,187,209]; 13 [13,143,169]

# Cramer coin-flip model, matched to the wheel's prime density
import math, random
random.seed(7)
density = 210 / 48                       # = 4.375; primes are this much denser
def model_prime(n):
    return random.random() < min(1.0, density / math.log(n))
nums = [210*k + r for k in range(51) for r in units if 210*k + r > 1]
print(sum(model_prime(n) for n in nums))   # 1356 (seed 7)
```

Chapter 2

The Mirror-Pair Populations

On the mod-210 wheel the reflection $r \mapsto 210 - r$ sorts the 48 coprime spokes into 24 mirror pairs, and on any given turn each pair falls into one of three classes by the primality of its two members: prime–prime, prime–subprime, or subprime–subprime. This chapter records a single, deliberately modest fact about how those classes are populated. Turn after turn, their counts are reproduced—to within per-turn noise, across the first six hundred turns—by the prime density alone: if p is the fraction of coprime spokes that are prime on a turn, the three classes hold $24p^2$, $48p(1-p)$, and $24(1-p)^2$ pairs. No correlation between the two partners, and no arithmetic beyond the density, is needed or detectable. We present this as an effective description of the texture, in the spirit of Cramér's model: the agreement is itself the content, because it shows the pair census carries nothing the prime number theorem does not already supply. It is not a new signal, a mechanism, or a spectrum, and we are explicit about that throughout.

2.1 Scope

This chapter is entirely classical and introduces no result. It quantifies one descriptive feature of the mod-210 wheel set up in the opening chapter: how the wheel's mirror pairs distribute among three primality classes as the wheel turns, and why that distribution is fixed by the prime density and nothing else [2]. The purpose is partly bookkeeping—to capture the pattern correctly, so that it is not mistaken for structure it does not have.

2.2 The three pair classes

The mirror $r \mapsto 210 - r$ splits the 48 coprime spokes into 24 pairs $\{r,\ 210 - r\}$. On turn k the pair occupies the integers $210k + r$ and $210k + (210 - r)$, and each is either prime or a product of primes ≥ 11 (a *subprime*); the lone exception, the unit 1, sits on turn 0 and is set aside. Each of the 24 pairs therefore belongs, on each turn, to exactly one of three classes:

$$\text{prime–prime (PP)}, \qquad \text{prime–subprime (PS)}, \qquad \text{subprime–subprime (SS)}. \tag{2.1}$$

On the first turn the primes are dense and PP dominates; as the wheel turns and the primes thin, PP gives way to SS, with PS in between. The question is whether anything governs that handover beyond the thinning itself.

2.3 The density model

Let $p = p(k)$ be the fraction of the 48 coprime spokes carrying a prime on turn k. Near $n = 210k$ the wheel's prime density is $1/\ln n$ lifted by the factor $210/\varphi(210) = 210/48$, so

$$p(k) \approx \frac{210/48}{\ln(210k)}. \tag{2.2}$$

If the two members of a mirror pair are each prime with probability p, *independently*—the Cramér-style heuristic [3], the partners sharing only their coprimality to 210—then the 24 pairs split binomially:

$$\langle \mathrm{PP} \rangle = 24\,p^2, \qquad \langle \mathrm{PS} \rangle = 48\,p(1-p), \qquad \langle \mathrm{SS} \rangle = 24\,(1-p)^2. \tag{2.3}$$

Figure 2.1 sets these forms against the counts read directly off the wheel. The left panel follows the three classes turn by turn: the binned counts track Eq. (2.3) as p falls from about 0.73 to 0.39 over six hundred turns, PP and SS crossing near $p = \frac{1}{2}$. The right panel removes the turn axis entirely and plots the same counts against p; they land on the three curves of Eq. (2.3) with only the scatter expected from 24 pairs per turn. The model has no free parameters.

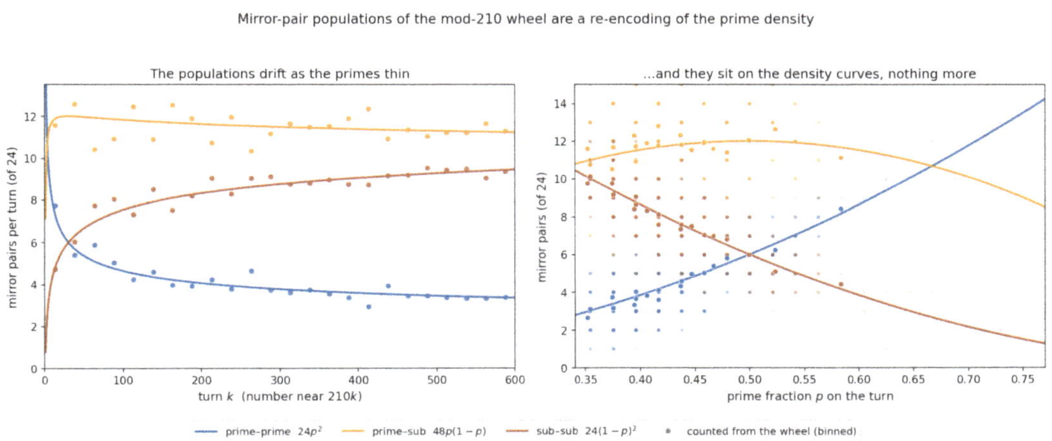

Figure 2.1: Mirror-pair populations of the wheel against the density model, Eq. (2.3). **Left:** the three classes versus turn k; lines are $24p^2$, $48p(1-p)$, $24(1-p)^2$ with $p = (210/48)/\ln n$, dots are counted from the wheel and binned. As the primes thin, prime–prime pairs give way to subprime–subprime, with prime–subprime holding near its maximum. **Right:** the same counts plotted against the prime fraction p on each turn (faint points per turn, solid points binned) fall onto the three curves. The populations are a re-encoding of the prime density; there is no further structure.

2.4 What the agreement means

The fit is the point, and it cuts against reading anything more into the classes. That the counts match Eq. (2.3) to within the per-turn scatter says the two members of a mirror pair behave as independent draws at the local density: there is no detectable correlation between a residue's primality and its mirror's, beyond the coprimality both already carry. Whatever "damping" distinguishes the three classes—the steady decline of PP, the rise of SS—is therefore not a property of the pairing at all. It is the prime number theorem, the $1/\ln n$ thinning, seen through the mirror. Naming a separate damping for each class would only be renaming $p(k)$ three times.

This places the pair census firmly on the descriptive side of the ledger. It is an effective model of texture, of exactly the kind Cramér's coin-flip is for the primes themselves [3]: faithful to the statistics, silent on the mechanism. It reproduces no zeros, defines no operator, and isolates no signal; it cannot, because it contains only the density, and the density is the smooth backbone the primes and subprimes already share. In the language of the diffraction chapter, the subprimes are the periodic clockwork and the primes the aperiodic remainder, but this particular statistic sees neither the clockwork's comb nor the remainder's zeros—only the slowly falling envelope common to both. That is why it is so well behaved, and why it is not, on its own, telling us anything the prime number theorem has not already told us.

A final word on status. The independence in Eq. (2.3) is a heuristic, not a theorem; it is of a piece with the Hardy–Littlewood conjectures on correlations among primes in residue classes, which remain open. The agreement here is a finite-window observation over the first six hundred turns, offered as a clean description and nothing stronger.

Counting the classes

The three populations need no heuristic to *count*—only the wheel and a primality test:

```
from sympy import isprime
from math import gcd
units = [r for r in range(1, 210) if gcd(r, 210) == 1]
pairs = sorted({tuple(sorted((r, 210 - r))) for r in units})   # 24 mirror pairs
def classes(k):                              # (PP, PS, SS) on turn k
    PP = PS = SS = 0
    for a, b in pairs:
        pa, pb = isprime(210*k + a), isprime(210*k + b)
        if pa and pb: PP += 1
        elif pa or pb: PS += 1
        else:          SS += 1
    return PP, PS, SS
print(classes(1), classes(100), classes(300))   # (13, 9, 2) (7, 10, 7) (3, 9, 12)
```

Part II

Distribution on the spokes

Chapter 3

Equidistribution

This expository chapter records the zeroth-order law of the primes on the wheel modulo 210*: the law that the later chapters measure deviations from. Two classical theorems do the work. Dirichlet's theorem (1837) guarantees that each of the* 48 *residue classes coprime to* 210 *contains infinitely many primes; its engine is the non-vanishing* $L(1, \chi) \neq 0$ *of the Dirichlet L-functions. The prime number theorem for arithmetic progressions then makes the stronger statement that the primes are* equidistributed *among those* 48 *classes — each receives, asymptotically, the same share* 1/48*. We illustrate this directly: the primes up to* 10^7 *fall across the* 48 *classes with a spread of only* 0.26%*, and that spread shrinks as the cutoff grows. The small residual wobble around exact equality is precisely where Chebyshev's prime race lives. Everything here is classical; this chapter simply fixes the leading-order picture before the race chapter studies the correction.*

3.1 Scope

This chapter is expository and entirely classical [4, 2, 5]. It states the two results—Dirichlet's theorem and the prime number theorem for arithmetic progressions—that govern the coarse distribution of primes across the wheel, and it shows them numerically. It is the foundation the prime-race chapter builds on: that chapter's entire subject, Chebyshev's bias, is the second-order fluctuation around the equidistribution established here. The opening chapter describes the wheel itself.

3.2 The forty-eight rays

Before the formulas, a picture. Fix a coprime residue r and follow the spoke it sits on through every turn of the wheel: the numbers r, $r + 210$, $r + 420, \ldots$ all share the same angle and march outward, so over infinitely many turns the spoke fills in to a solid ray running from the centre to the rim. There are exactly 48 such rays, one per coprime residue, and a ray is nothing other than an arithmetic progression $\{ 210k + r : k \geq 0 \}$ drawn radially (Fig. 3.1). The rim, reached only in the limit, plays the role of infinity.

This single image carries the two theorems of this chapter in advance. Dirichlet's theorem says no ray is ever starved: each of the 48 holds infinitely many primes. The prime number theorem for progressions says more — the rays are *balanced*, each collecting the same limiting share 1/48 of all primes — so the boundary measure the rays induce at infinity is simply uniform on 48 points. The geometry also makes the wheel's mirror literal. The map $r \mapsto 210 - r$ is reflection across the vertical diameter; it carries each ray to its partner and sorts the 48 into 24 mirror pairs (the horizontal rungs in the figure). Whether a Dirichlet character is symmetric or antisymmetric under that reflection is exactly its parity $\chi(-1) = \pm 1$, the organizing distinction taken up in the conductor-lattice chapter.

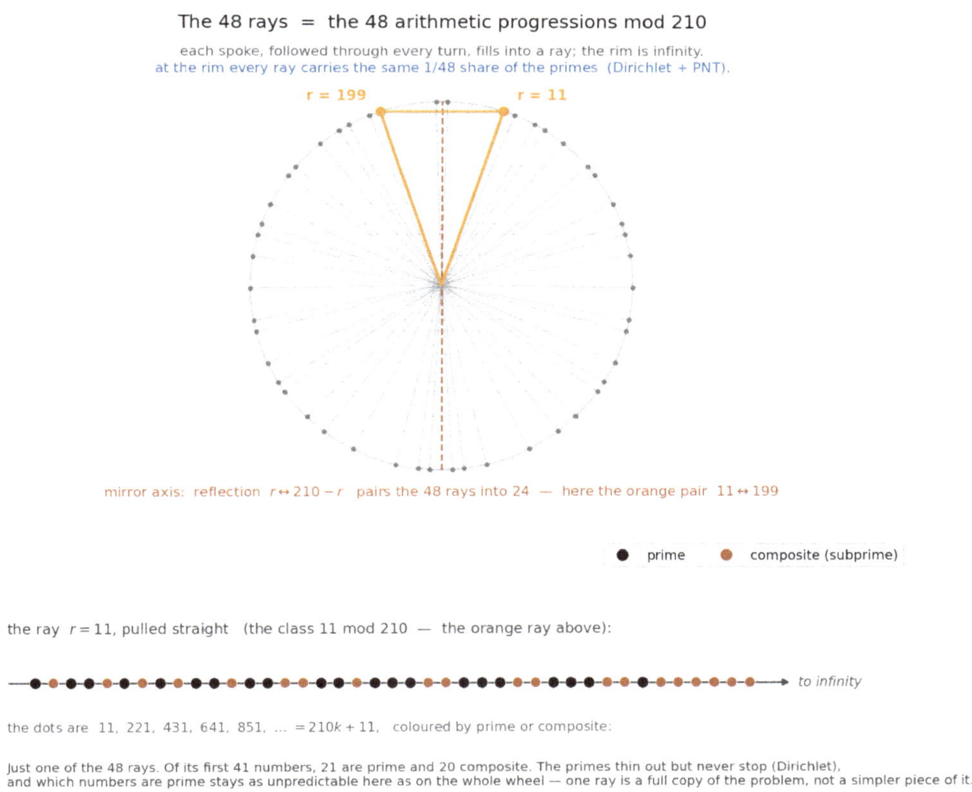

Figure 3.1: Top: the 48 rays, one per coprime residue, with the 24 horizontal rungs marking the mirror pairs $r \leftrightarrow 210 - r$ and the vertical dashed line the axis of that reflection. At the rim — the limit of infinitely many turns — each ray carries the same 1/48 share of the primes. Bottom: a single ray unrolled, the class 11 mod 210 (the numbers $11, 221, 431, \cdots = 210k + 11$), with its primes (black) and subprimes (red) over the first 41 numbers. Zooming into one ray does not simplify anything: the primes along it are as pseudorandom as on the whole wheel, only sparser.

Two cautions keep the picture honest. First, the 48 rays are an organizing device, not a simplification: each ray is a full-size copy of the same hard problem, carrying a thinned-out but equally pseudorandom stream of primes, so no single ray is the easy one. Second, the natural way to work with the rays is not one at a time but through the Dirichlet characters mod 210 — specific weighted blends of all the rays that diagonalize the wheel's symmetry. A single character touches every ray at once, and recovering one ray means summing over all the characters by orthogonality; this is precisely the mechanism Dirichlet's proof uses, and the subject of the next sections. (The "boundary at infinity" here is a faithful metaphor for the limit, not a hidden geometry: the deep boundary-at-infinity of number theory, where modular forms live, is a separate and far more structured construction.)

3.3 Dirichlet's theorem

The wheel sorts every prime $p > 7$ into one of the 48 residue classes coprime to 210. The first question is whether any class is ever starved of primes. Dirichlet answered it in 1837: for any a with $\gcd(a, m) = 1$,

$$\#\{\, p \text{ prime} : p \equiv a \pmod{m} \,\} = \infty. \tag{3.1}$$

Applied to $m = 210$, every one of the 48 coprime classes holds infinitely many primes; none runs dry.

The proof introduced the objects that organize the entire subject. To each character χ of $(\mathbb{Z}/m\mathbb{Z})^\times$ one attaches the L-function $L(s, \chi) = \sum_{n \geq 1} \chi(n) n^{-s}$, and the primes in a class are isolated by averaging these over χ. The argument succeeds because, for every non-principal χ,

$$L(1, \chi) \neq 0. \tag{3.2}$$

This single non-vanishing is the load-bearing fact: were any $L(1, \chi)$ to vanish, the primes could pile up away from some class and leave it finite. The characters of the wheel, and the seven real ones in particular, are the subject of the conductor-lattice chapter; here we need only that their L-functions stay clear of zero at $s = 1$.

3.4 Equidistribution

Dirichlet's theorem says every class is infinite; it does not yet say the classes are *balanced*. The prime number theorem for arithmetic progressions, established by de la Vallée Poussin in 1896 — the prime number theorem itself having been proved that year independently by Hadamard and de la Vallée Poussin — supplies the balance. For $\gcd(a, m) = 1$,

$$\pi(x; m, a) \sim \frac{1}{\varphi(m)} \frac{x}{\log x} \sim \frac{\pi(x)}{\varphi(m)}, \tag{3.3}$$

so each coprime class captures the *same* asymptotic fraction $1/\varphi(m)$ of all primes. On the wheel, $\varphi(210) = 48$, and every one of the 48 classes should hold about $\pi(x)/48$ of the primes up to x.

Figure 3.2 shows this for $x = 10^7$. There are $\pi(10^7) = 664{,}579$ primes; discarding $2, 3, 5, 7$ leaves 664,575 on the wheel, an average of 13,845 per class. The actual counts run from a minimum of 13,757 to a maximum of 13,922 — a coefficient of variation of just 0.26% about the mean. To the eye the 48 bars are simply level: equidistribution is not a subtle limiting statement here but a visible fact already at 10^7.

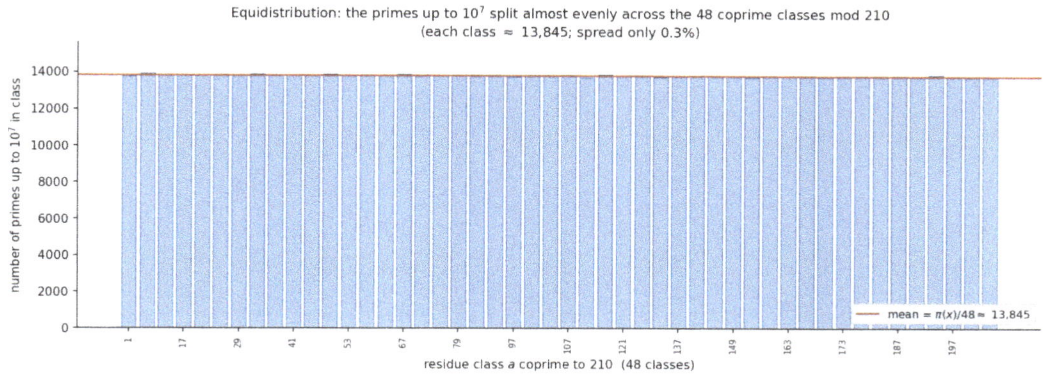

Figure 3.2: The primes up to 10^7 distributed across the 48 residue classes coprime to 210. Each bar is a class; the red line is the mean $\pi(x)/48 \approx 13{,}845$. The counts vary by only 0.26% from class to class — the equidistribution promised by Eq. (3.3), plainly visible.

3.5 The balance is asymptotic — and that is where the race lives

Equation (3.3) is a statement about the limit, and the approach to it is gradual. Measuring the spread across the 48 classes by their coefficient of variation (standard deviation over mean), one finds it falling steadily as the cutoff grows (Fig. 3.3):

$$6.5\% \;\; (x = 10^4) \;\longrightarrow\; 2.5\% \;\; (10^5) \;\longrightarrow\; 0.8\% \;\; (10^6) \;\longrightarrow\; 0.26\% \;\; (10^7). \qquad (3.4)$$

The classes are pulling level, but they are never exactly level at finite x. That residual imbalance is not noise to be discarded; it is the entire subject of the next chapter. When one resolves the small deviations $\pi(x; m, a) - \pi(x)/\varphi(m)$ they are found to be *systematic*: classes of one arithmetic type tend to lead, classes of another to lag. That persistent tilt is Chebyshev's bias, and its fine structure is controlled by the zeros of the same L-functions whose non-vanishing at $s = 1$ gave us Dirichlet's theorem in the first place [6].

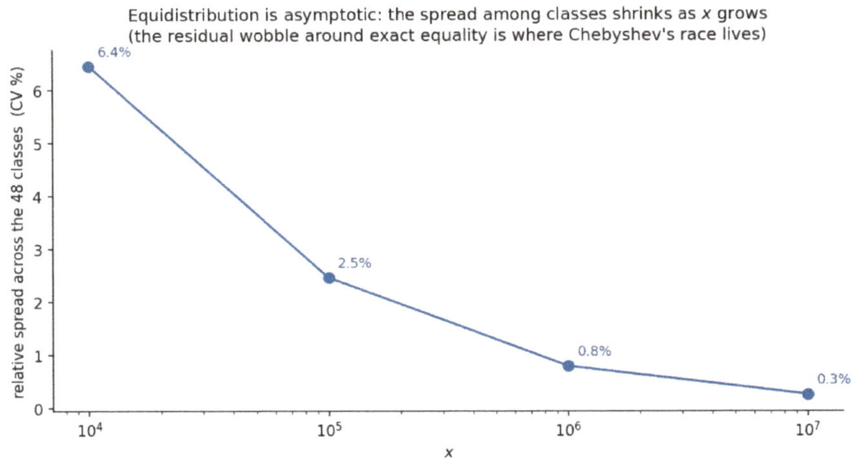

Figure 3.3: The relative spread among the 48 classes (coefficient of variation) as a function of the cutoff x. It shrinks toward zero, confirming that equidistribution is an asymptotic law. The small wobble that remains at any finite x is the deviation the prime-race chapter measures.

3.6 What this sets up

The picture to carry forward is a two-term one. To leading order the primes are blind to which spoke they land on: all 48 classes are equal, by Eq. (3.3). The interesting arithmetic is in the correction. The race chapter measures the systematic part of that correction; the conductor chapter studies the characters that sort the classes into the competing teams; and the diffraction chapter traces the correction back to the L-function zeros. All three presuppose the flat zeroth-order law established here.

Reproducing the figures

```
import numpy as np
from math import gcd

N = 10_000_000
sieve = np.ones(N+1, bool); sieve[:2] = False
for i in range(2, int(N**0.5)+1):
    if sieve[i]: sieve[i*i::i] = False
primes = np.nonzero(sieve)[0]

units = [r for r in range(210) if gcd(r, 210) == 1]   # the 48 classes
pp = primes[primes > 7]

# Figure 1: counts per class at x = 1e7
res = pp % 210
counts = np.array([np.count_nonzero(res == r) for r in units])
print(counts.sum(), counts.mean(), counts.min(), counts.max(),
```

```
      100*counts.std()/counts.mean())        # 664575 13845.3 13757 13922 0.26

# Figure 2: spread (CV%) vs cutoff
for X in [1e4, 1e5, 1e6, 1e7]:
    c = np.array([np.count_nonzero((pp[pp <= X]) % 210 == r) for r in units], float)
    print(X, round(100*c.std()/c.mean(), 2))   # 6.45, 2.46, 0.79, 0.26

# Figure 3: one ray unrolled -- the class 11 mod 210 over its first 41 numbers
from sympy import isprime
ray = [(k, isprime(210*k + 11)) for k in range(41)]
nprime = sum(p for _, p in ray)
print(nprime, 41 - nprime)                     # 21 primes, 20 subprimes
```

Chapter 4

The Prime Race

This is an expository *chapter. We make a classical phenomenon visible by computing it directly on a single small modulus: we sort primes into two competing families of residue classes modulo* 210*, count them block by block, and watch the race. The lead changes hands endlessly, the cumulative advantage wanders at the characteristic* $\sqrt{N}$ *rate rather than accumulating, the prime "population" of each block thins as the local form of the prime number theorem, and—underneath all of it—the fluctuations are governed by the zeros of Dirichlet L-functions. None of this is new: the bias was first noticed by Chebyshev (1853), the modern framework is due to Rubinstein and Sarnak (1994), and a celebrated accessible account is Granville and Martin's* Prime Number Races *(2006). For this particular (mirror-symmetric) split we also pin down the leading direction of the bias—the lower team is favored—and isolate the one genuinely remaining computation, the exact logarithmic density from the L-function zeros, honestly as exactly that. The aim here is only clarity—to let a reader* watch *the zeros act through an elementary count—and to mark honestly where the small experiment connects to the deepest open question in the subject.*

4.1 Scope

This chapter contains no new theorem and makes no claim outside number theory. The computations are elementary sieving; the explanations are textbook. Its value, if any, is pedagogical. A reader who wants the rigorous machinery behind every observation below should consult Refs. [6, 7, 2, 5].

4.2 The wheel

Fix the modulus $M = 210 = 2 \cdot 3 \cdot 5 \cdot 7$, with the $\varphi(210) = 48$ live spokes and the mirror involution $r \mapsto 210 - r$ of the opening chapter (Fig. 4.1). Every integer n lands on spoke $n \bmod 210$; advancing by one *period* (one block $\mathcal{B}_k = [210k, 210k + 210)$) returns to the same spoke. The mirror pairs each spoke with another ($11 \leftrightarrow 199$, etc.) and exchanges the lower semicircle $(0, 105)$ with the upper $(105, 210)$, each half holding 24 live spokes.

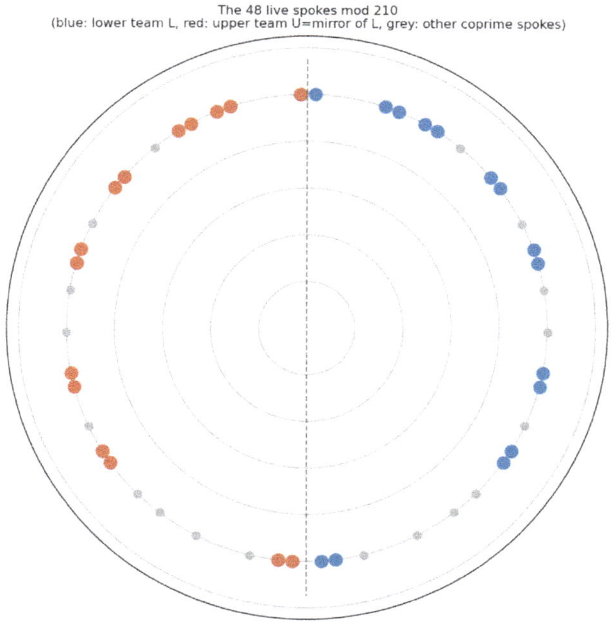

Figure 4.1: The 48 live spokes modulo 210. Blue: the lower team $\mathcal{L}$; red: the upper team $\mathcal{U}$, its mirror image; grey: the remaining coprime spokes. The two teams are reflections of each other across the 0–105 axis (dashed).

4.3 Two complementary teams

Among the live spokes, single out the *twin-admissible* ones: residues r for which r and $r+2$ are *both* coprime to 210 (the only positions where a twin-prime pair can sit). Split these by the mirror into two equal teams:

$$\mathcal{L} = \{1, 11, 13, 17, 19, 29, 31, 41, 43, 59, 61, 71, 73, 101, 103\}, \tag{4.1}$$

$$\mathcal{U} = \{107, 109, 137, 139, 149, 151, 167, 169, 179, 181, 191, 193, 197, 199, 209\}, \tag{4.2}$$

fifteen spokes each, with $\mathcal{U} = \{210 - r : r \in \mathcal{L}\}$. Most appear in visible twin pairs $(11, 13), (17, 19), (29, 31), \ldots,$ and one pair wraps the seam: spoke 1 is twin-admissible because 209 and $211 \equiv 1$ are both coprime, so the twin $(419, 421)$ straddles a block boundary. Nothing is special about "twin" spokes here; they merely give one clean, symmetric way to form two equal teams, and any balanced split behaves identically.

4.4 The race

For each period k, count the primes landing on each team,

$$A_{\mathcal{L}}(k) = \#\{p \text{ prime} : p \in \mathcal{B}_k,\ p \bmod 210 \in \mathcal{L}\}, \qquad A_{\mathcal{U}}(k) \text{ likewise}, \tag{4.3}$$

and form the per-period asymmetry $\Delta(k) = A_{\mathcal{L}}(k) - A_{\mathcal{U}}(k)$. Does either team build a durable lead?

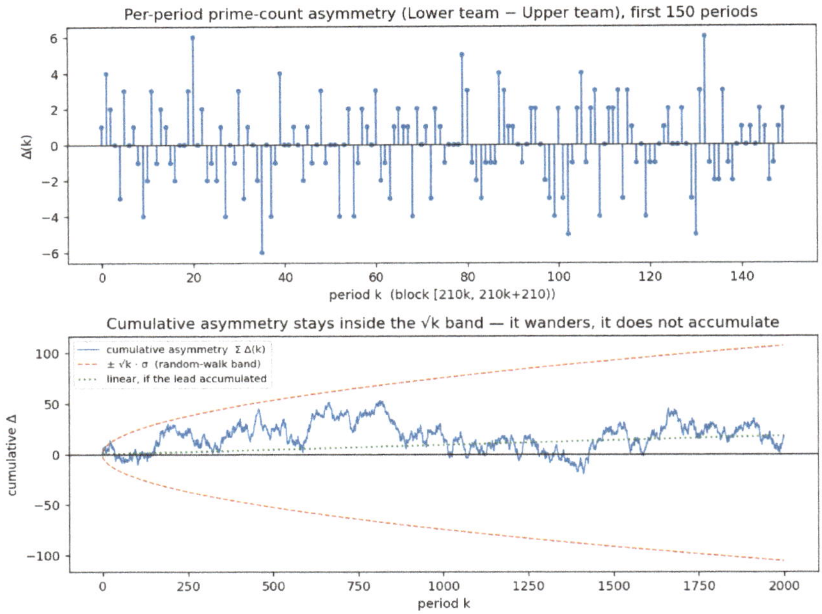

Figure 4.2: Top: the per-period lead $\Delta(k)$ oscillating around zero (first 150 periods). Bottom: the cumulative lead stays inside the $\pm\sqrt{k}\,\sigma$ random-walk band (red dashed) and nowhere near the straight line a genuinely accumulating bias would follow (green dotted).

4.5 What the race does

Computed over the first 2000 periods (integers up to $\sim 4.2 \times 10^5$), the answer is vivid (Figs. 4.2 and 4.3):

- **The lead flips constantly.** $\Delta(k)$ changes sign every few periods; over 500 periods there are about 208 sign changes, with $\mathcal{L}$ ahead in 213, $\mathcal{U}$ ahead in 195, and 92 ties. The per-period spread is small, $\sigma \approx 2.4$.

- **The cumulative lead does not run away.** $\sum_k \Delta(k)$ is +26 after 500 periods and +18 after 2000—it *shrinks* back toward zero with more data. Against the random-walk envelope $\pm\sqrt{N}\,\sigma$ (± 53 at $N{=}500$, ± 106 at $N{=}2000$), the cumulative stays comfortably inside the band throughout.

- **No net bias is detectable here.** The cumulative is consistent with an unbiased $\sqrt{N}$ wander. Whether a genuine, tiny bias exists for *this* team split is exactly the question the Rubinstein–Sarnak analysis settles (it depends on how many spokes in each team are quadratic residues mod $3, 5, 7$); raw counting at this scale cannot resolve it directly, and Sec. 4.9 extracts the direction analytically instead.

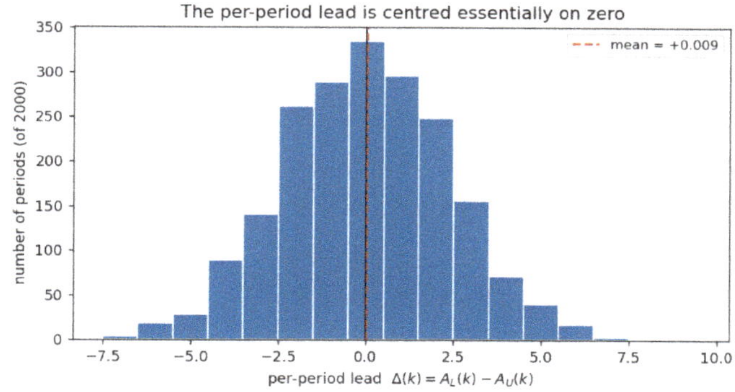

Figure 4.3: Distribution of the per-period lead $\Delta(k)$ over 2000 periods: centred essentially on zero (mean +0.009).

4.6 The thinning: the prime number theorem on one wheel

A second, slower motion runs alongside the race. In each block the 48 live spokes split between *primes* and *subprimes* (composites coprime to 210, i.e. with smallest prime factor ≥ 11). In the first block only five subprimes fit—$121, 143, 169, 187, 209$—because the smallest possible one is $11^2 = 121$; that block is 42 primes, 5 subprimes, and the unit. As k grows, primes thin and subprimes occupy the freed spokes (Fig. 4.4). Measured directly: the prime count per block tracks $210/\log(210k)$ with correlation $\approx$0.55—simply the prime number theorem read off block by block, the residual being the same fluctuation as in the race; the two reach parity (24 each) near $k = 19$ ($N \approx 4000$); and the mean changeover rate, fit over the first $\sim$110 periods, is about +0.077 subprimes per period. This "interchange" is a smooth classical thinning buried in arithmetic fluctuation, carrying no structure beyond $\pi(x) \sim x/\log x$. The fill-in is moreover *uniform across the wheel*: at $X = 2 \times 10^6$ the subprime-to-prime ratio is essentially constant over all 48 classes—mean 2.07, coefficient of variation under 1% (range 2.04–2.11)—and the two teams receive equal shares (2.06 versus 2.07). The ratio itself grows slowly with X as composites accumulate, but it grows the *same* in every class, so the subprimes obey the same equidistribution the primes do (Sec. 4.7), filling the freed spokes evenly rather than favoring either team.

4.7 Why the race cancels: equidistribution

Both motions trace to one classical fact, established in the equidistribution chapter: the primes are *equidistributed* across the 48 classes coprime to 210, each receiving its fair share $1/\varphi(210)$ in the limit. Two equal teams therefore receive equal shares, so the team totals balance and $\sum_k \Delta(k)$ has no linear drift. The lead cannot run away because the field is, in the limit, perfectly level. What remains is the *rate of approach* to that level field—the fluctuation—and that is where the depth lives.

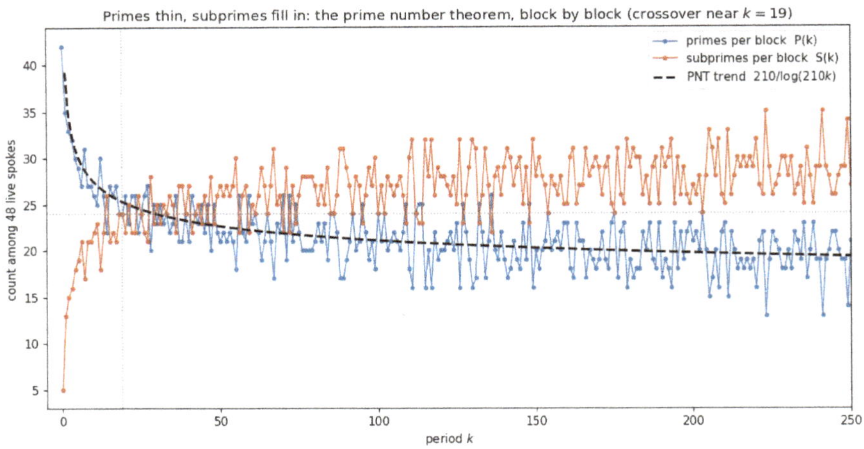

Figure 4.4: Primes thin, subprimes fill in, block by block. The dashed curve is the prime-number-theorem trend $210/\log(210k)$; the two populations cross near $k = 19$.

4.8 Why it can only wander: the zeros

The fluctuation is not arbitrary. By the explicit formula—the bridge Riemann opened in 1859 [9], in its form for arithmetic progressions—the error in counting primes in a residue class is a sum over the nontrivial zeros $\rho = \frac{1}{2} + i\gamma$ of the relevant Dirichlet L-functions,

$$(\text{prime count in class}) - (\text{expected}) \;\approx\; -\sum_{\rho} \frac{x^{\rho}}{\rho}. \tag{4.4}$$

Each zero contributes an oscillation; their sum is the wobble in $\Delta(k)$. Two consequences seen directly above: under the Riemann Hypothesis every $\operatorname{Re}\rho = \frac{1}{2}$, so the error is of order $\sqrt{x}$—precisely the $\sqrt{N}$ band the cumulative race lived inside; and because infinitely many zeros keep contributing, the lead changes hands forever, which is the content of the Chebyshev-bias / Rubinstein–Sarnak analysis [6] and its limiting logarithmic density. The small experiment on the 210 wheel is, underneath, a direct sighting of the zeros of L-functions—the same objects whose spacing statistics famously resemble random-matrix eigenvalues [10] and whose location is the Riemann Hypothesis. That connection is the genuine prize, and it is a question *in mathematics*.

4.9 The leading bias: which way the race tilts

The direct count could not name a winner—at this scale the tilt is buried under the $\sqrt{N}$ fluctuations—but the Rubinstein–Sarnak framework [6] extracts it anyway, and the mirror symmetry of our teams makes the leading term clean.

First, a simplification. Because $\mathcal{U} = \{210 - r : r \in \mathcal{L}\}$, every *even* Dirichlet character ($\chi(-1) = +1$) contributes equally to both teams and cancels in the difference.

The race between two mirror-image teams is therefore governed entirely by the *odd* characters modulo 210.

The bias itself is the Chebyshev effect: residue classes that are perfect squares mod q are prime-deficient (the squares of primes accumulate there), so a team carrying more quadratic residues wins less often. Modulo 210 there are exactly six coprime quadratic residues, $\{1, 79, 109, 121, 151, 169\}$, each with eight square roots. Of these the lower team contains only one (the residue 1), while the upper team contains three $(109, 151, 169)$:

$$\sum_{a\in\mathcal{L}} c(a) = 8, \qquad \sum_{a\in\mathcal{U}} c(a) = 24, \qquad c(a) = \#\{t \bmod 210 : t^2 \equiv a\}. \tag{4.5}$$

The upper team is the more prime-deficient, so the race is *not* fair: the lower team is favored, and its Rubinstein–Sarnak logarithmic density of leading exceeds $\frac{1}{2}$. This agrees in direction with the finite-X observation that the lower twin-corridor half is the "Chebyshev-favored" half, and with the block-0 lean of 14 primes to 13 seen above.

Two honest caveats close the loop. First, this persistent leading bias coexists with the count ratio tending to 1 (equidistribution): the cumulative totals equalize while one side still leads more than half the time—that coexistence *is* the Chebyshev paradox. Second, the *direction* ($\mathcal{L}$ favored) is now settled, but the *exact* density requires the variance contributed by the nontrivial zeros of the odd Dirichlet L-functions modulo 210. That is the one genuinely remaining computation—a known method applied to a non-canonical example, conditional on GRH and the linear-independence hypothesis—and we leave it open here as exactly that.

4.10 What this is and is not

This chapter re-derives by hand, on one wheel, the prime-race / Chebyshev-bias picture. The findings—oscillating lead, $\sqrt{N}$ wander, prime-number-theorem thinning, control by the zeros—are textbook consequences of theory that is 30 to 170 years old. It is not a new method or result, and makes no claim outside number theory. Its only ambition is to let a reader watch the zeros act through a simple count. The rigorous next step—computing the exact Rubinstein–Sarnak bias density for a chosen team split—is a real, specialized calculation within the existing framework [6]. The broader study of unexpected biases in the distribution of primes remains active; for a striking modern instance in a different guise—biases between *consecutive* primes—see Lemke Oliver and Soundararajan [8].

Reproducing the computation

```
import numpy as np, math
M = 210
Z  = [r for r in range(1, M) if math.gcd(r, M) == 1]    # 48 live spokes
Zs = set(Z); T = set()
for r in Z:                                              # twin-admissible spokes
    if (r + 2) % M in Zs:
```

```
        T.add(r); T.add((r + 2) % M)
L = sorted(r for r in T if r < M/2)                   # lower team (15)
U = sorted((M - r) % M for r in L)                    # upper team (15)

Nper = 2000; top = M*(Nper + 1)
spf = np.zeros(top + 1, dtype=np.int64)               # smallest-prime-factor sieve
for i in range(2, top + 1):
    if spf[i] == 0:
        spf[i::i] = np.where(spf[i::i] == 0, i, spf[i::i])
isprime = lambda n: n >= 2 and spf[n] == n

d = np.array([sum(isprime(M*k+r) for r in L) - sum(isprime(M*k+r) for r in U)
             for k in range(Nper)])
C = np.cumsum(d)
print("cumulative:", int(C[-1]), " sqrt-band:", round(np.sqrt(Nper)*d.std(), 1))
```

The leading bias term (Sec. 4.9) and the subprime equidistribution (Sec. 4.7) reproduce as follows.

```
# leading bias term: quadratic residues mod 210 (each has 8 square roots)
QR = {(x * x) % M for x in Z}
c  = lambda a: 8 if a in QR else 0
print("sum c over L:", sum(c(a) for a in L),     # 8  (one QR: residue 1)
      " over U:", sum(c(a) for a in U))          # 24 (three QRs: 109, 151, 169)

# subprime equidistribution: per-class subprime/prime ratio at X = 2e6
X = 2_000_000
s = np.zeros(X + 1, dtype=np.int64)
for i in range(2, X + 1):
    if s[i] == 0:
        s[i::i] = np.where(s[i::i] == 0, i, s[i::i])
ratio = []
for r in Z:
    start = r if r > 1 else r + M                # skip the unit at n = 1
    prime = sum(1 for n in range(start, X + 1, M) if s[n] == n)
    sub   = sum(1 for n in range(start, X + 1, M) if n > 1 and s[n] != n)
    ratio.append(sub / prime)
ratio = np.array(ratio)
print("ratio mean %.3f  CV %.4f" % (ratio.mean(), ratio.std() / ratio.mean()))
```

Chapter 5

The Conductor Lattice

This chapter is an expository *extension of the diffraction chapter, "Seeing the Primes in the Zeros," which drew the Riemann zeros as a quasicrystal whose diffraction is the primes, and sketched the same picture for two characters of the mod-210 wheel. Here we carry it out for all seven* real *Dirichlet characters modulo* 210 *and make the organizing structure explicit. The seven characters are the nontrivial divisors of* $105 = 3 \cdot 5 \cdot 7$*, which form a Boolean lattice; each is a Kronecker symbol attached to a fundamental discriminant; each L-function's zeros form a quasicrystal on the same prime lattice; and the conductor of a character switches off exactly its own peaks among* $\{\log 3, \log 5, \log 7\}$*, leaving* $\log 2$ *and every prime from* 11 *upward always present. The even/odd grading of the characters—the wheel's mirror* $r \mapsto 210 - r$*—turns out to be precisely the parity of* $|c \cap \{3, 7\}|$*. Everything here is classical and fully explicit: nothing beyond Kronecker symbols and the explicit formula is required, and no new result is claimed.*

5.1 Scope

This chapter proves nothing new. Its content is standard Dirichlet–Kronecker theory [2, 5], drawn concretely on a single modulus and paired with the diffraction picture from the chapter on the zeros. The value, if any, is that every statement is exact and every figure is computed from the actual L-function zeros. A reader who wants the general theory should consult Refs. [2, 5].

5.2 The seven real characters

The Dirichlet characters modulo 210 form a group dual to $(\mathbb{Z}/210\mathbb{Z})^\times$. Among them, the *real* characters are those of order dividing two—the quadratic characters—and the nontrivial real *primitive* ones are exactly the Kronecker symbols

$$\chi_D(\cdot) = \left(\frac{D}{\cdot}\right), \tag{5.1}$$

Table 5.1: The seven real primitive characters modulo 210. "Parity" is $\chi(-1)$; "dark primes" are the primes $p \mid c$, for which $\chi(p) = 0$ and the corresponding diffraction peak is absent.

χ	conductor c	D	symbol	$\chi(-1)$	$\lvert c \cap \{3,7\}\rvert$	dark primes
$\chi_{3.2}$	3	-3	$(-3/\cdot)$	-1 (odd)	1	$\log 3$
$\chi_{5.4}$	5	$+5$	$(5/\cdot)$	$+1$ (even)	0	$\log 5$
$\chi_{7.6}$	7	-7	$(-7/\cdot)$	-1 (odd)	1	$\log 7$
$\chi_{15.14}$	15	-15	$(-15/\cdot)$	-1 (odd)	1	$\log 3,\ \log 5$
$\chi_{21.20}$	21	$+21$	$(21/\cdot)$	$+1$ (even)	2	$\log 3,\ \log 7$
$\chi_{35.34}$	35	-35	$(-35/\cdot)$	-1 (odd)	1	$\log 5,\ \log 7$
$\chi_{105.104}$	105	$+105$	$(105/\cdot)$	$+1$ (even)	2	$\log 3,\ \log 5,\ \log 7$

where D runs over the fundamental discriminants whose conductor divides 210. Because 2 does not ramify here, these conductors are the nontrivial divisors of $105 = 3 \cdot 5 \cdot 7$, namely

$$c \in \{3,\ 5,\ 7,\ 15,\ 21,\ 35,\ 105\}. \tag{5.2}$$

Each prime contributes a *prime discriminant*

$$p^* := (-1)^{(p-1)/2}\, p, \qquad \text{so} \qquad 3^* = -3, \quad 5^* = +5, \quad 7^* = -7, \tag{5.3}$$

the sign being $-$ for $p \equiv 3 \pmod 4$ and $+$ for $p \equiv 1 \pmod 4$. The discriminant of a composite conductor is the product of the prime discriminants of its prime factors, e.g. $15 \to (-3)(5) = -15$ and $21 \to (-3)(-7) = +21$. Table 5.1 lists the seven characters, their discriminants, parities, and the primes each one extinguishes.

5.3 The conductor lattice

The seven conductors are the nontrivial divisors of 105, and under divisibility they form the Boolean lattice on the set $\{3, 5, 7\}$—the subset lattice, drawn as a cube in Fig. 5.1. The bottom vertex is the trivial conductor 1 (the principal character, not one of the seven); the three atoms are the prime conductors; the three middle vertices are the products of two primes; the top is 105 itself.

The parity of each character reads straight off the lattice. Since $\chi_D(-1) = \operatorname{sgn}(D)$, and D is the product of the prime discriminants $\{3^*, 5^*, 7^*\} = \{-3, +5, -7\}$, the sign of D flips once for each prime $\equiv 3 \pmod 4$ dividing the conductor—that is, once for each of $\{3, 7\}$ it contains. Hence

$$\chi \text{ is even} \iff D > 0 \iff |c \cap \{3,7\}| \text{ is even}. \tag{5.4}$$

This gives the even set $\{5, 21, 105\}$ and the odd set $\{3, 7, 15, 35\}$ (three even, four odd), shown by colour in Fig. 5.1. The same even/odd split is the wheel's mirror symmetry $r \mapsto 210 - r$: even characters are fixed by it, odd characters are negated by it. The grading that in the prime-race chapter decides which characters carry the bias is, here, the parity of a two-element intersection.

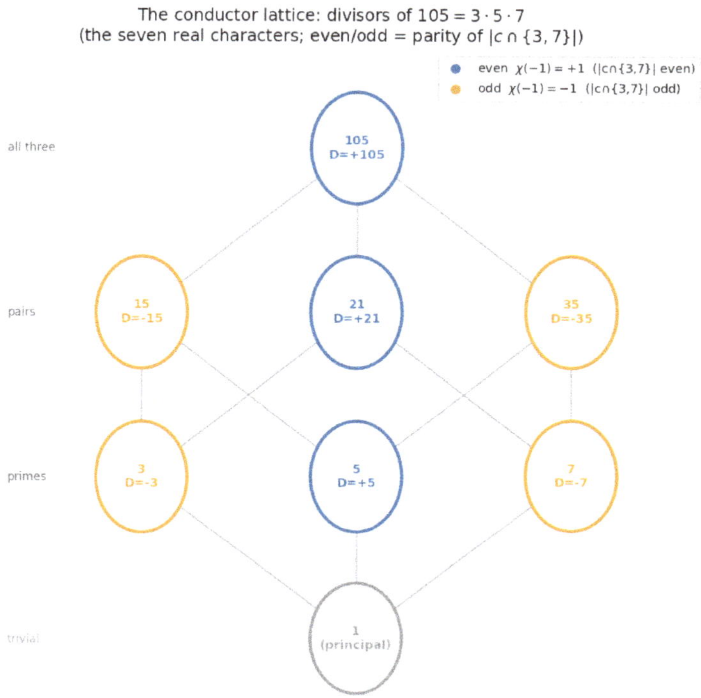

Figure 5.1: The conductor lattice: the divisors of $105 = 3 \cdot 5 \cdot 7$ as a Boolean cube. The seven nontrivial divisors are the seven real characters; colour denotes parity, which equals the parity of $|c \cap \{3, 7\}|$. The trivial conductor 1 (principal character) sits at the bottom.

5.4 Each character as a quasicrystal

Each real character χ of conductor c has an L-function $L(s, \chi)$, and the Guinand–Weil explicit formula for $L(s, \chi)$ is again a Fourier duality between its nontrivial zeros $\rho = \frac{1}{2} + i\gamma_\chi$ and the primes: for a suitable test function,

$$\sum_{\gamma_\chi} h(\gamma_\chi) \;=\; (\text{smooth terms}) \;-\; \sum_p \sum_{m \geq 1} \frac{\chi(p^m) \log p}{p^{m/2}} \, \widehat{h}(m \log p) \,. \tag{5.5}$$

The right-hand side differs from the ζ case only by the weight $\chi(p^m)$. So the diffraction of the zeros of $L(s, \chi)$,

$$D_\chi(u) = \sum_{\gamma_\chi} w(\gamma_\chi) \cos(\gamma_\chi u), \tag{5.6}$$

peaks at the same positions $u = \log(p^m)$ as before, but the sign of each peak is the Kronecker symbol $\chi(p) = \pm 1$, and a prime $p \mid c$ contributes nothing at all, because $\chi(p) = 0$.

Two facts make the picture clean (Fig. 5.2). First, 2 divides none of the conductors, so the $\log 2$ peak is always present. Second, no prime ≥ 11 divides 105, so every peak from $\log 11$ upward is always present. The only peaks the conductor can switch off are therefore those of $\{3, 5, 7\}$—and it switches off exactly the ones dividing c. Climbing the lattice extinguishes more of $\{\log 3, \log 5, \log 7\}$, until at $\chi_{105.104}$ all three are gone.

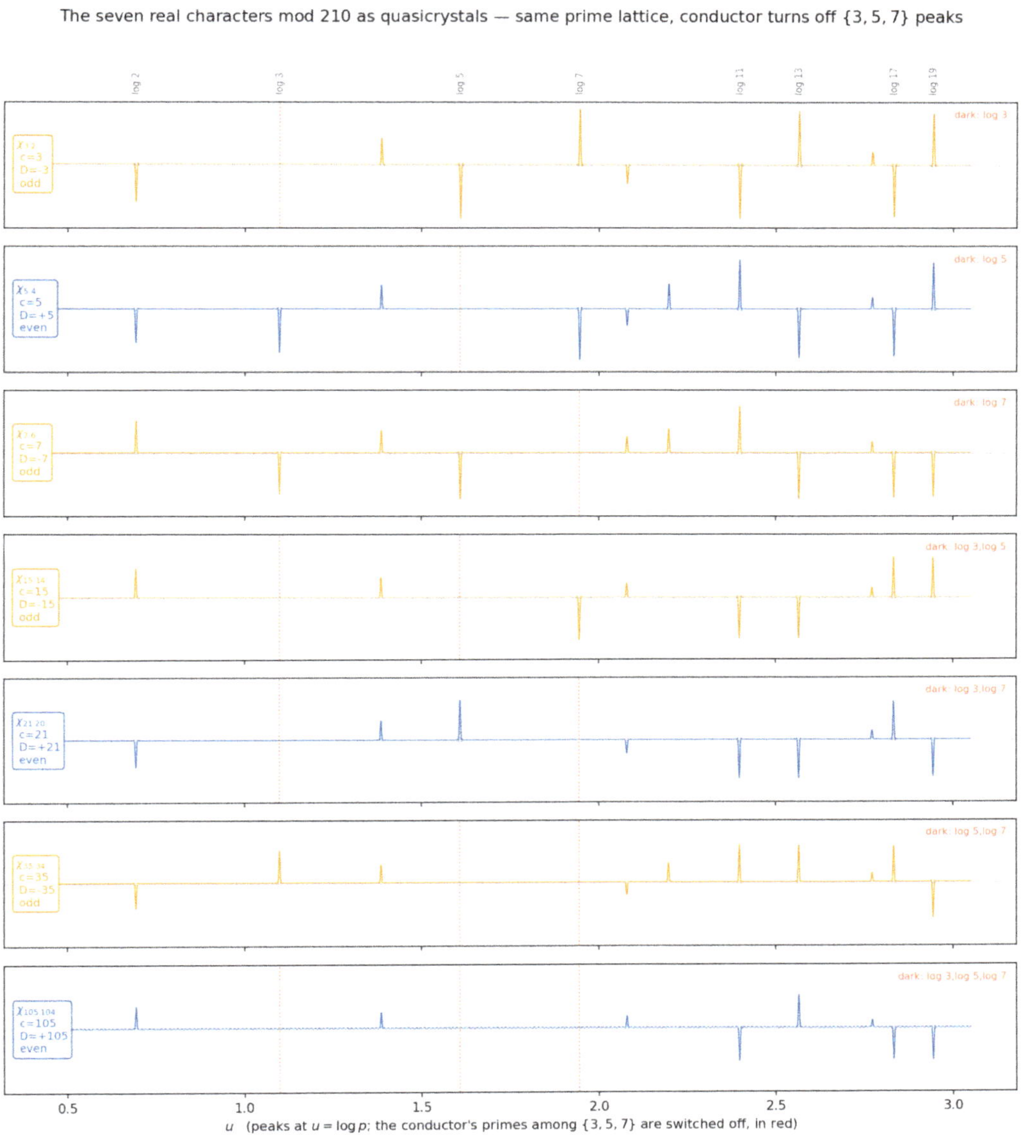

Figure 5.2: The seven real characters modulo 210, each drawn as the diffraction of its L-function zeros (computed from the actual zeros up to height $T = 1500$). All share one prime lattice; the sign of each live peak is $\chi(p)$. Red dotted lines mark the peaks switched off by the conductor (the primes $p \mid c$). Reading top to bottom, the conductor climbs the lattice and extinguishes more of $\{\log 3, \log 5, \log 7\}$; $\chi_{105.104}$ (bottom) is the maximally filtered member, with $\log 3$, $\log 5$, and $\log 7$ all dark and only $\log 2$ and the primes from 11 up surviving.

5.5 The family

Put together, Figs. 5.1 and 5.2 say the same thing two ways. The seven characters are one family of quasicrystals built on a single prime lattice; what distinguishes them is only (i) which of $\{3, 5, 7\}$ is extinguished—a position on the Boolean cube—and (ii) the sign pattern $\chi(p)$ on the surviving primes. The vertical axis of the cube is a filtration: the higher the conductor climbs, the more of $\{3, 5, 7\}$ goes dark, with $\chi_{105.104}$ at the top stripped down to $\log 2$ and the primes ≥ 11. The colouring across the cube is the parity, equivalently the wheel's mirror. Nothing in the family escapes the prime lattice underneath; the conductor only chooses which spokes of it to dim.

Two of these seven are the repeating-decimal families. The chapter on the wheel's repeating decimals sorts the 48 residues into two *families* according to the periodic block of $r/210$. That split is not an independent structure: it is one of the seven characters tabulated above, and *which* one is fixed entirely by the base in which the wheel is written. Writing the wheel in base 10, the family of a residue r is the value of the Legendre symbol $\left(\frac{r}{3}\right)$—that is, the conductor-3 character $\chi_{3.2}$. Writing it in base 60 (a clock's base), the family is $\left(\frac{r}{7}\right)$—the conductor-7 character $\chi_{7.6}$. The mechanism is the coset arithmetic of the repeating period: base 10's powers generate all of $(\mathbb{Z}/7)^\times$ but fix $(\mathbb{Z}/3)^\times$ (since $10 \equiv 1 \pmod 3$), so $\langle 10\rangle$ is the kernel of $\left(\frac{\cdot}{3}\right)$ and the family is $\left(\frac{\cdot}{3}\right)$; base 60 instead has $3 \mid 60$ (genuinely absorbed) and $\langle 60\rangle = \{1, 2, 4\}$, the squares modulo 7 — the kernel of $\left(\frac{\cdot}{7}\right)$ — so the family is $\left(\frac{\cdot}{7}\right)$. Both correspondences are exact—every base-10 period block consists of residues sharing a single value of $\left(\frac{\cdot}{3}\right)$, and every base-60 block a single value of $\left(\frac{\cdot}{7}\right)$. So the two "families" of the decimals chapter are precisely the two lowest-conductor members of this lattice, $\chi_{3.2}$ and $\chi_{7.6}$, surfaced one at a time by the choice of base: a reminder that the digits are a lens and these characters are the object underneath.

5.6 What this is and is not

Everything above is classical and explicit. The real characters as Kronecker symbols, the fundamental discriminants, the conductor as a Boolean lattice of divisors, and the explicit formula (5.5) are textbook [2, 5]; the diffraction reading is the one from the diffraction chapter, itself a rendering of the explicit formula [9, 11, 12] in the quasicrystal language of Dyson [13]. This chapter simply organizes the seven and computes their diffractions.

It is worth being plain about what the conductor structure does and does not do. It *organizes* the family beautifully—it is the exact dial that says which of $\{3, 5, 7\}$ each character dims, and its parity is the mirror. But it is not a hidden key. A direct per-character study of these same seven characters finds no clean prime-versus-composite or conductor-magnitude signal; the only load-bearing distinction across the family is the parity, which is the mirror we already have. Computing the zeros of any one of these L-functions is a few seconds of standard software; the open problem is the analytic one, and it is the Generalized Riemann Hypothesis, one L-function at a time. The conductor lattice arranges the seven quasicrystals on a shelf; it does not crack any of them. And the bright peaks under all seven are, still, the primes.

Reproducing the figures

The diffraction of any character is the same cosine sum used in the diffraction chapter, applied to that character's zeros:

```
import numpy as np
def diffraction(gamma, u):                    # gamma: imaginary parts of L(s,chi) zeros
    g = gamma[gamma > 0]
    w = 0.5*(1 + np.cos(np.pi*g/g.max()))  # Hann taper
    return -(w[:,None]*np.cos(np.outer(g, u))).sum(0)/len(g)

u = np.linspace(0.45, 3.05, 3500)             # peaks appear at u = log(p^m)
```

The character table regenerates from Kronecker symbols. With `D` the fundamental discriminant and `p` an odd prime, $\chi_D(p)$ is the Kronecker symbol (D/p), equal to $+1, -1$, or 0; a character is even iff $D > 0$, and the dark primes are those dividing the conductor. The seven discriminants are $\{-3, +5, -7, -15, +21, -35, +105\}$, each the product of the prime discriminants $\{-3, +5, -7\}$ of the primes in its conductor.

Part III

Mirrors and the spectrum

Chapter 6

The Two Mirrors

The reflection $r \mapsto 210 - r$ has run through every chapter so far as a plain geometric fold of the wheel. This chapter connects it to a second, deeper reflection: the $s \mapsto 1 - s$ symmetry of the functional equation. The hinge between them is the parity of a character, $\chi(-1) = \pm 1$. The wheel's mirror sorts the characters into even (their values symmetric across the axis) and odd (antisymmetric); that very parity chooses the gamma factor, $\Gamma(s/2)$ or $\Gamma(\frac{s+1}{2})$, that completes $L(s, \chi)$ and fixes the shape of its functional equation, including where its trivial zeros fall. The two mirrors are the same fact wearing two costumes. Everything here is classical—Riemann's functional equation for ζ and its extension to Dirichlet L-functions; this chapter only places the wheel's reflection beside the analytic one.

6.1 Scope

This chapter is expository and classical [9, 2, 5]. It states the functional equation for the wheel's L-functions and shows how the geometric mirror of the opening chapter and the even/odd grading of the conductor-lattice chapter are tied to it through one quantity, the parity $\chi(-1)$. It introduces no result; the functional equation is a theorem more than a century and a half old. The Riemann Hypothesis, which concerns the one feature of the analytic mirror this chapter does *not* settle, is flagged in Sec. 6.5 and otherwise left to the diffraction chapter.

6.2 The wheel's mirror and the parity of a character

The wheel's reflection is $r \mapsto 210 - r \equiv -r \pmod{210}$. A Dirichlet character interacts with it in the simplest possible way:

$$\chi(210 - r) = \chi(-r) = \chi(-1)\,\chi(r), \tag{6.1}$$

so the reflection multiplies every value by the single sign $\chi(-1)$. A character is called *even* if $\chi(-1) = +1$ and *odd* if $\chi(-1) = -1$, and on the wheel the names are literal: an even character takes equal values at r and its mirror $210 - r$, an odd one takes opposite values. Figure 6.1 shows both, the quadratic character modulo 5 (even) and modulo 3 (odd): the even wheel is symmetric in color across the axis, the odd wheel is color-reversed.

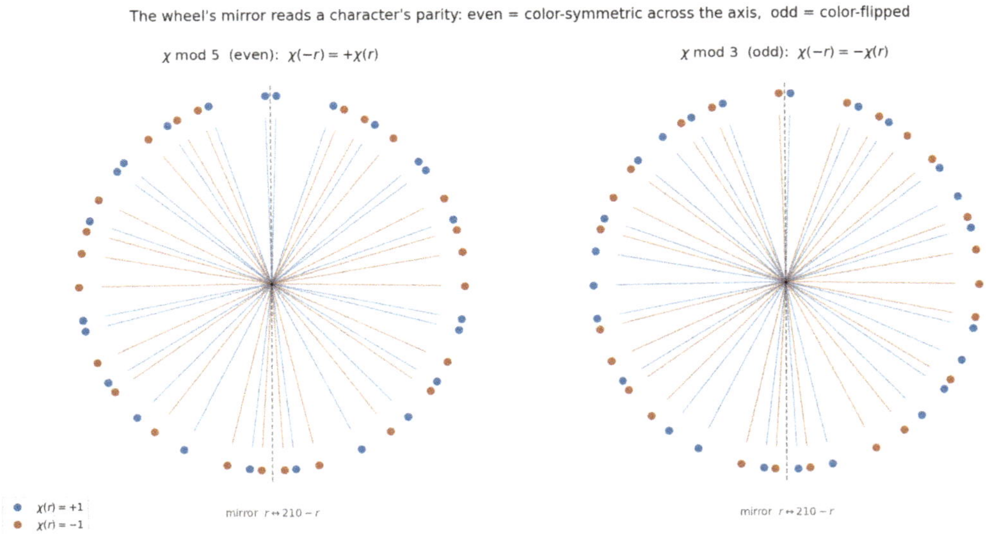

Figure 6.1: The wheel's mirror reading a character's parity. Left: the even character χ mod 5, whose values satisfy $\chi(-r) = +\chi(r)$, so the coloring is symmetric across the dashed mirror axis. Right: the odd character χ mod 3, with $\chi(-r) = -\chi(r)$, so the coloring flips across the axis. Blue marks $\chi(r) = +1$ and red $\chi(r) = -1$.

For the seven real characters of the wheel this sorts them three-and-four. The even ones are the characters of conductor $5, 21, 105$ (fundamental discriminant $D > 0$); the odd ones have conductor $3, 7, 15, 35$ ($D < 0$). The parity is exactly the even/odd grading the conductor-lattice chapter found from the discriminant sign: the wheel's mirror and the sign of D are the same partition.

6.3 The functional equation and its mirror

The deeper reflection lives not on the wheel but in the complex plane of the variable s. For a primitive character χ modulo q, set $a = 0$ if χ is even and $a = 1$ if χ is odd—that is, $\chi(-1) = (-1)^a$—and form the completed L-function

$$\Lambda(s,\chi) = \left(\frac{q}{\pi}\right)^{(s+a)/2} \Gamma\!\left(\frac{s+a}{2}\right) L(s,\chi). \tag{6.2}$$

It satisfies the functional equation

$$\Lambda(s,\chi) = \varepsilon(\chi)\,\Lambda(1-s,\overline{\chi}), \qquad |\varepsilon(\chi)| = 1, \tag{6.3}$$

relating the value at s to the value at its mirror image $1-s$. The reflection $s \mapsto 1-s$ is a point reflection through $s = \frac{1}{2}$, which leaves the critical line $\mathrm{Re}(s) = \frac{1}{2}$ invariant — the axis of the symmetry, and the mirror line of the s-plane (Fig. 6.2).

The parity enters Eq. (6.2) through the single integer a, and it is decisive. An even character ($a = 0$) is completed by $\Gamma(s/2)$; an odd one ($a = 1$) by $\Gamma(\frac{s+1}{2})$. Because the completed function is entire and non-vanishing away from the nontrivial zeros, $L(s, \chi)$ must vanish wherever its gamma factor has a pole, and those poles sit at different places for the two parities:

$$\text{even: trivial zeros at } s = 0, -2, -4, \ldots; \qquad \text{odd: trivial zeros at } s = -1, -3, -5, \ldots \tag{6.4}$$

The Riemann zeta function is the even prototype: $\xi(s) = \pi^{-s/2}\Gamma(s/2)\zeta(s)$ (with a factor $\frac{1}{2}s(s-1)$ to absorb the pole) obeys $\xi(s) = \xi(1-s)$, and ζ's trivial zeros are the even ones, $-2, -4, \ldots$ So the choice of gamma factor, made by the parity, sets both the precise form of the $s \mapsto 1-s$ symmetry and the location of the trivial zeros on the negative real axis.

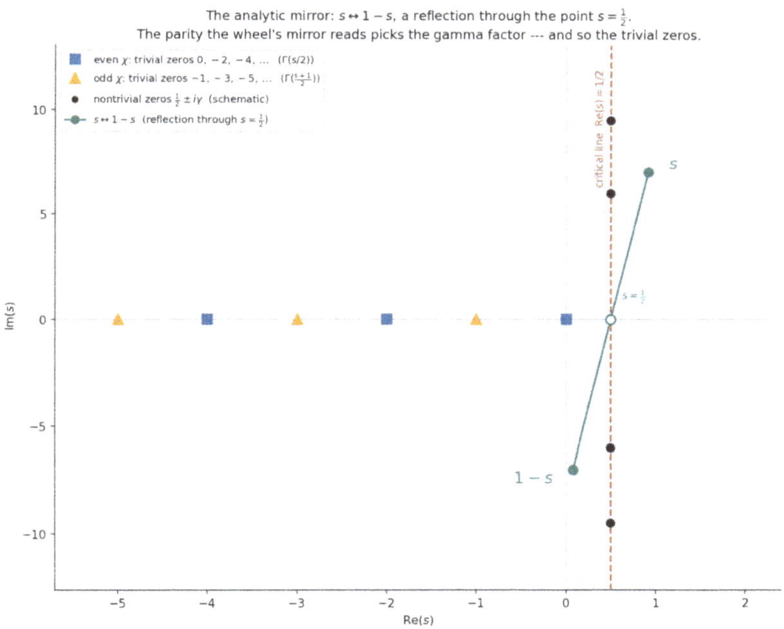

Figure 6.2: The analytic mirror. The functional equation (6.3) sends a point s to its image $1-s$ (green), a point reflection through $s = \frac{1}{2}$ on the critical line $\mathrm{Re}(s) = \frac{1}{2}$ (red). Even characters are completed by $\Gamma(s/2)$ and carry trivial zeros at $0, -2, -4, \ldots$ (blue squares); odd characters by $\Gamma(\frac{s+1}{2})$, with trivial zeros at $-1, -3, -5, \ldots$ (orange triangles). The nontrivial zeros (black, schematic) lie in the critical strip; the Riemann Hypothesis is the conjecture that they lie exactly on the mirror axis.

6.4 The two mirrors are one

The chain is short and it closes. The wheel's reflection $r \mapsto 210 - r$ acts on a character as multiplication by $\chi(-1)$; that sign is the parity; the parity fixes the integer a; a

selects the gamma factor $\Gamma(\frac{s+a}{2})$ in Eq. (6.2); and that gamma factor determines the exact functional equation (6.3) under $s \mapsto 1-s$. The geometric fold on the wheel and the analytic fold in the s-plane are therefore not two coincidental symmetries but two readings of one quantity. When the opening chapter named the mirror "a fact about residues, not factorization," this is where the fact reappears: the same reflection that recolors the wheel decides the shape of the L-function's reflection.

6.5 What this is and is not

The functional equation is a theorem, proved by Riemann for ζ in 1859 [9] and extended to the Dirichlet L-functions by the classical methods recorded in any text [2, 5]. The parity dependence of the gamma factor is equally standard. The contribution of this chapter is only the framing: to set the wheel's reflection beside the analytic one and identify the hinge. The "two mirrors" is a pedagogical packaging of a real correspondence, not a new mathematical statement.

One feature of the analytic mirror is emphatically not settled here. The functional equation places the nontrivial zeros symmetrically about the critical line; it does not place them *on* it. That the zeros lie exactly on the mirror axis $\mathrm{Re}(s) = \frac{1}{2}$ is the Riemann Hypothesis, open for ζ and for every L-function on the wheel. This chapter establishes the symmetry of the mirror, not the location of what it reflects.

Parity of the seven real characters

```
# A real character mod 210 has conductor a divisor of 105 = 3*5*7.
# Even iff |conductor cap {3,7}| is even  (i.e. fundamental disc D > 0).
for cond in [{3},{5},{7},{3,5},{3,7},{5,7},{3,5,7}]:
    parity = "even" if len(cond & {3,7}) % 2 == 0 else "odd"
    print(sorted(cond), parity)
# {3} odd   {5} even   {7} odd   {3,5} odd
# {3,7} even   {5,7} odd   {3,5,7} even
# -> even: conductors 5, 21, 105 ;  odd: conductors 3, 7, 15, 35
```

Chapter 7

Seeing the Primes in the Zeros

This is an expository *chapter paired with the prime-race chapter, "A Prime Race on the Mod-210 Wheel." Where that chapter watched the primes compete in the* counting *domain—who is ahead, by how much—this one looks at the same objects in the* spectral *domain. Riemann's explicit formula is a Fourier duality between the nontrivial zeros of $\zeta(s)$ and the primes. We make it visible the most direct way possible: we Fourier-transform a bare list of the first thousand zeros and watch sharp peaks appear at exactly the logarithms of the primes and prime powers. Viewed this way the zeros are, conditionally on the Riemann Hypothesis, a one-dimensional* quasicrystal *whose diffraction pattern is the primes. We then draw the same picture for the Dirichlet characters of the mod-210 wheel: each character's L-function gives its own quasicrystal on the same prime lattice, with the sign of each peak fixed by $\chi(p)$, and the wheel's mirror $r \mapsto 210 - r$ appearing as the even/odd grading of the characters. None of this is new. The duality is the explicit formula (Riemann 1859; Guinand; Weil); the metaphor is the "music of the primes" (Berry; Mazur–Stein); the quasicrystal framing is Dyson's (2009). The only ambition is to* see *it, concretely, on one wheel.*

7.1 Scope

This chapter contains no new theorem and makes no claim outside number theory. The picture it draws is one of the most reproduced in the subject. Its value, if any, is pedagogical: to let a reader watch a classical duality happen on real data. A reader who wants the rigorous statements should consult Refs. [9, 11, 12, 2, 5]. The single piece here that is a fresh *rendering* rather than a textbook classic is Sec. 7.6, where the characters of one specific modulus are drawn side by side; even there the *structure* is entirely standard.

7.2 The duality

Riemann's explicit formula, in the form due to Guinand and Weil [11, 12], equates a sum over the nontrivial zeros $\rho = \frac{1}{2} + i\gamma$ of $\zeta(s)$ with a sum over the primes:

schematically, for a suitable test function h with Fourier transform $\widehat{h}$,

$$\sum_{\gamma} h(\gamma) = (\text{smooth archimedean terms}) - \sum_{p} \sum_{m \geq 1} \frac{\log p}{p^{m/2}} \widehat{h}(m \log p) . \tag{7.1}$$

The right-hand side is supported on the numbers $\log(p^m)$. So the zeros and the primes are a Fourier pair: each is, in the language of harmonic analysis, the spectrum of the other. This is the exact sense of the popular metaphor that the zeros carry the *music of the primes* [14, 15]; each zero $\rho = \frac{1}{2} + i\gamma$ contributes to the prime distribution an oscillation $x^\rho = \sqrt{x}\, e^{i\gamma \log x}$ of frequency γ.

7.3 The same primes, in real space

The duality just stated—zeros and primes as a Fourier pair—has two concrete pictures, one in each domain. This section draws the spatial one; the rest of the chapter draws the spectral one.

Lay the integers on the mod-210 wheel and let it turn, so that the number $210k + r$ sits at residue r on turn k. Each prime $p \geq 11$ then marks the multiples it owns on a single family of evenly spaced parallel lines—a *grating* of slope $-(210 \bmod p)$, striking a given spoke once every p turns (Fig. 7.1). Three gratings already resemble an optical element; laid over one another (Fig. 7.2) they make moiré interference, and the clear channels no line crosses are the spokes where the primes survive. It is the sieve of Eratosthenes, drawn as overlapping diffraction gratings.

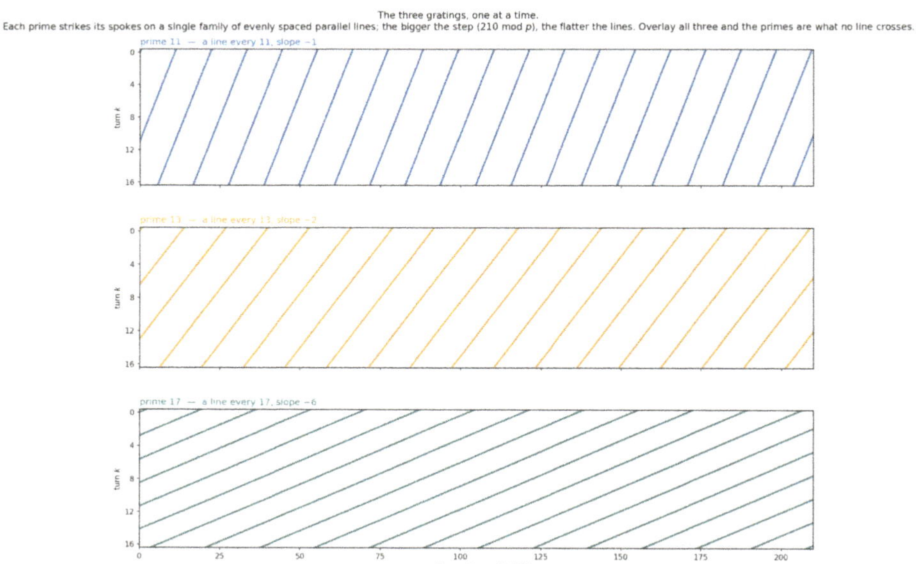

Figure 7.1: The three smallest gratings of the wheel, one per panel. Each prime p marks its multiples on a single family of parallel lines, spaced p apart and sloped by $-(210 \bmod p)$; the larger that step, the flatter the lines. These are the periodic beats of the sieve, shown separately before being overlaid.

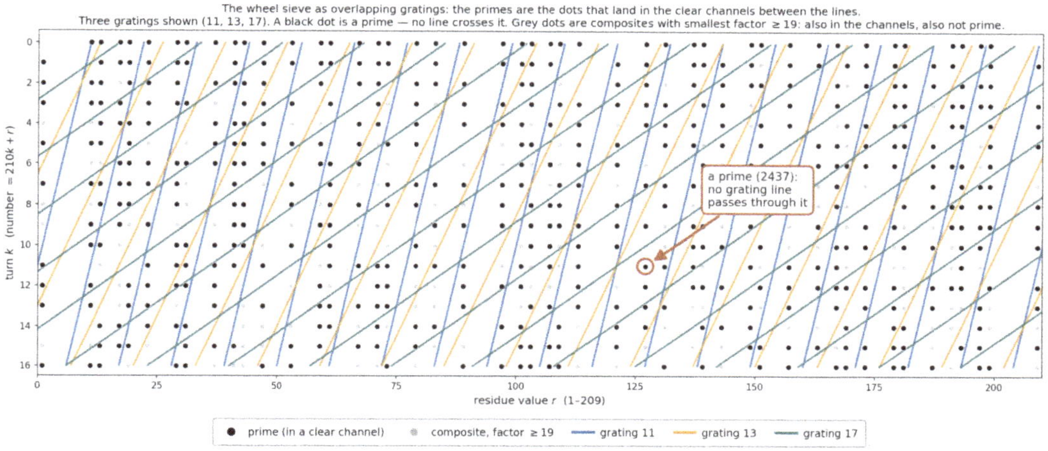

Figure 7.2: The same three gratings overlaid: the wheel's sieve as moiré interference. A black dot is a prime—a spoke no line crosses. Grey dots are composites whose smallest factor is ≥ 19: also in the open channels, also not prime, caught only by gratings not drawn here. A true prime survives the gratings of *every* prime up to $\sqrt{n}$.

This real-space picture carries a single honest seam—the one the whole chapter circles. Every grating is perfectly periodic, and any finite stack of them repeats with the period of a primorial: the composites are clockwork an optical bench could reproduce. The primes are what remains once all that periodicity is removed—the channels surviving every grating up to $\sqrt{n}$—and that remainder has no period at all. In real space, physics grips the gratings completely and what they leave behind not at all.

By Eq. (7.1), the spectrum of that same prime distribution is the set of zeros, so the diffraction picture of the next section (Fig. 7.3) shows these very primes from the spectral side. The two are companion renderings, not literal transforms of one another—the overlay is an analogy for the sieve, the explicit formula is the exact duality—but together they make one point: the periodic part of the prime story is the part that behaves like physics, and the primes themselves are the aperiodic remainder.

7.4 The picture

Equation (7.1) can be run as an experiment. Take only the imaginary parts of the first $N = 1000$ zeros—a list of bare numbers $14.13, 21.02, 25.01, \ldots$ with no reference to primes anywhere in it—and form the cosine sum

$$D(u) = \sum_{n=1}^{N} w_n \cos(\gamma_n u), \qquad (7.2)$$

where w_n is a smooth taper that suppresses truncation ringing. Plotted against u (Fig. 7.3, top), $D(u)$ is flat and small almost everywhere and erupts into sharp peaks at exactly $u = \log 2,\ \log 3,\ \log 5,\ \log 7,\ \log 11, \ldots$, with smaller peaks at the prime powers $\log 4$, $\log 8$, $\log 9$. The peak heights follow $\log p\,/\,p^{m/2}$ from Eq. (7.1), largest around $p \sim 5$–13 and decaying thereafter. The primes were not put in; they fall out of the zeros.

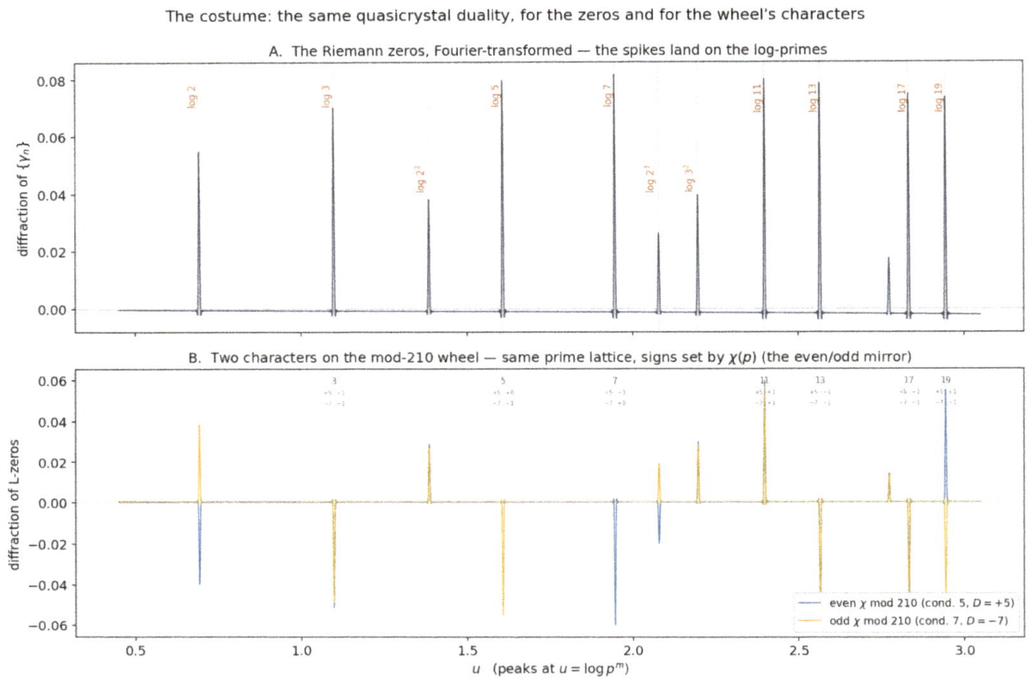

Figure 7.3: **Top:** the Fourier transform (7.2) of the first 1000 zeros of $\zeta(s)$. The peaks land on the logarithms of the primes and prime powers (dotted red). The zeros, transformed, are the primes. **Bottom:** the same construction for two Dirichlet characters of the mod-210 wheel, computed from their L-function zeros—one even (χ of conductor 5, $D = +5$) and one odd (χ of conductor 7, $D = -7$). The peaks sit on the same prime lattice, but each peak's *sign* is $\chi(p)$; a prime dividing the conductor goes dark. The even/odd grading is the wheel's mirror $r \mapsto 210 - r$.

7.5 The quasicrystal

A one-dimensional quasicrystal, in Dyson's sense [13], is a discrete set of points on a line whose Fourier transform is *also* a discrete set of points. Figure 7.3 is exactly that statement: a discrete set of points (the zeros, lying on the critical line) whose diffraction is a discrete set of points (the log-primes). Quasicrystals are real materials—Shechtman's diffraction patterns with crystallographically forbidden symmetry won a Nobel Prize—so this is the most literal sense in which the Riemann spectrum could be said to appear "in nature": a hypothetical one-dimensional atomic arrangement whose X-ray diffraction would show bright spots at the zeros.

Two honest qualifications. The peaks-at-log-primes of Fig. 7.3 are *unconditional*—they are Eq. (7.1), true whether or not the Riemann Hypothesis holds. What requires the Hypothesis is the word *quasicrystal* itself: to call the zeros a pure point set lying *on a single line* is to assert $\operatorname{Re}\rho = \frac{1}{2}$ for all of them. So the diffraction is solid ground; the crystallographic name is the conditional dressing. And the picture, beautiful as it is, settles nothing: it is a faithful redescription of the primes, not an independent object that pins them. Dyson's proposal—classify all one-dimensional quasicrystals and find the zeros among them—remains a program, because the quasicrystal whose diffraction is the log-primes simply *is* the primes.

7.6 The wheel's characters

The mod-210 wheel enters through its Dirichlet characters. Each character χ modulo 210 has an L-function $L(s,\chi)$, and each L-function carries its own explicit formula—a version of Eq. (7.1) in which the prime p is weighted by $\chi(p)$. So the zeros of $L(s,\chi)$ form their own quasicrystal, on the *same* prime lattice as before, but with each peak's sign set by $\chi(p)$. Figure 7.3 (bottom) draws two of them, computed from their L-function zeros: an even character (conductor 5) and an odd one (conductor 7). Where $\chi(p) = +1$ the peak points with ζ; where $\chi(p) = -1$ it flips; where p divides the conductor the peak vanishes entirely. The grading of the characters into even $(\chi(-1) = +1)$ and odd $(\chi(-1) = -1)$ is exactly the wheel's mirror $r \mapsto 210 - r$—the same mirror that, in the prime-race chapter, sends the contest onto the odd characters. The wheel does not light up its own spokes; it indexes a *family* of quasicrystals, one per character, all sharing the prime lattice and differing only by a character's sign pattern.

7.7 What this is and is not

The phenomenon in Fig. 7.3 is classical and has been drawn many times: as the "music of the primes" [14, 15], as Fourier-duality plots of the zeros against the primes [16], and as a standalone computation of the same transform [17]. The quasicrystal reading is Dyson's [13]. The spacing statistics of the zeros that make them look like a physical spectrum—the random-matrix universality shared with quantum-chaotic systems—are Montgomery's and Rubinstein–Sarnak's territory [10, 6], not pursued here. This chapter adds none of that; it only renders the duality on one modulus and notes, in passing, that the wheel's characters give a family of these crystals. It bears on the Riemann Hypothesis not at all: the diffraction is a consequence of the primes, and the bright spots are the primes. It is the primes, drawn in crystal.

Reproducing the picture

```
import numpy as np
from mpmath import zetazero                    # generate the zeros (no data file)
gamma = np.array([float(zetazero(k).imag) for k in range(1, 1001)])
g = gamma[gamma > 0]
w = 0.5*(1 + np.cos(np.pi*g/g.max()))              # Hann taper -> clean peaks
u = np.linspace(0.45, 3.05, 4000)
```

```
D = -(w[:,None]*np.cos(np.outer(g, u))).sum(0)/len(g)   # the diffraction, Eq. (2)

import math                                           # peaks should sit at log(p^m):
for p in [2,3,5,7,11,13,17,19]:
    print(p, round(math.log(p),3))                    # 0.693, 1.099, 1.609, 1.946, ...
```

For a character χ (Fig. 7.3, bottom), replace gamma with the imaginary parts of the zeros of $L(s, \chi)$; the peaks appear at the same $\log p$ with sign $\chi(p)$, no extra bookkeeping required.

The real-space companion (Figs. 7.1, 7.2) needs no zeros at all—each prime's grating is one line of modular arithmetic:

```
from math import gcd
units = [r for r in range(210) if gcd(r, 210) == 1]    # the 48 spokes
def strikes(p, r, k):                        # does prime p's grating hit spoke r on turn k?
    return (210*k + r) % p == 0               # i.e. p divides 210k + r
def open_channels(k, ps=(11, 13, 17)): # spokes no drawn grating crosses
    return [r for r in units if all(not strikes(p, r, k) for p in ps)]
print(len(open_channels(0)))                  # 40: unit 1 plus 39 primes, 19..199
```

Chapter 8

The Zeros as a Spectrum

The diffraction chapter read the zeros as the Fourier-dual of the primes. This chapter looks instead at the zeros among themselves. *Once their slowly growing density is removed, the nontrivial zeros of* ζ*—and of the seven L-functions of the wheel—carry the statistics of the eigenvalues of a large random Hermitian matrix from the Gaussian Unitary Ensemble: they repel at short range, almost never falling close together, and their pair correlation follows Montgomery's law* $1 - \left(\sin \pi u / \pi u\right)^2$*. For the first thousand zeros, no normalized gap falls below* 0.1*, and only* 7% *fall below* 0.5 *where an uncorrelated sequence would put* 39%*. These are the statistics of a quantized chaotic system, and they are the principal evidence for the Hilbert–Pólya idea that the zeros are the spectrum of some self-adjoint operator. The statistical agreement is established; the operator is not known, and the Riemann Hypothesis is not thereby proved. This chapter is expository.*

From here the tour borrows a vocabulary it has not used before. This chapter and the next read the wheel's zeros and primes through ideas from physics—random-matrix statistics here, a gas of particles in the chapter that follows—and the words can look like a change of subject. They are not. Every object stays one already built: the zeros of the wheel's L-functions, the explicit formula, the Möbius function. Only the lens changes. Where a term from physics is needed it is given a plain meaning before it is named, and the names are kept so they can be looked up; nothing here requires physics you do not meet on the page.

8.1 Scope

This chapter is expository and reports established mathematics [10, 18, 19]. It describes the local statistics of the zeros and their match to random-matrix theory, illustrates the match with the first thousand zeros of ζ, and states the quantum-chaos and Hilbert–Pólya context that the match suggests. It proves nothing. The one thing this circle of ideas would most like to deliver—an operator whose spectrum is the zeros, hence a proof of the Riemann Hypothesis—is exactly what remains open, and Sec. 8.6 is careful about that line. The diffraction chapter treats the complementary picture, the zeros against the primes.

8.2 Unfolding

The nontrivial zeros $\frac{1}{2} + i\gamma_n$ are not uniformly dense: near height T they crowd together with mean density $\frac{1}{2\pi}\log(T/2\pi)$, so the average gap shrinks as one climbs the critical line. To compare the zeros with a universal statistical law one first removes this trend—*unfolds* them—by applying the Riemann–von Mangoldt counting function

$$\widetilde{\gamma}_n = N(\gamma_n), \qquad N(T) = \frac{T}{2\pi}\log\frac{T}{2\pi} - \frac{T}{2\pi} + \frac{7}{8}, \tag{8.1}$$

which stretches the sequence to unit mean spacing. The unfolded gaps $s_n = \widetilde{\gamma}_{n+1} - \widetilde{\gamma}_n$ then have mean 1 (here, mean 1.000, standard deviation 0.38 over the first thousand zeros), and it is their *distribution* that carries the arithmetic.

8.3 Level repulsion

The first and most visible feature is that the zeros avoid one another. Figure 8.1A is the histogram of the unfolded gaps. It rises from zero, peaks near $s \approx 0.9$, and falls—there is almost no probability of a very small gap. This is *level repulsion*, and it is quantitative: among the first thousand zeros not a single normalized gap is below 0.1, and only 7.3% lie below 0.5. An uncorrelated (Poisson) sequence with the same density would crowd 39% of its gaps below 0.5 and pile up at $s = 0$ (the dashed curve e^{-s}). The data instead tracks the Gaussian Unitary Ensemble surmise

$$p_{\text{GUE}}(s) = \frac{32}{\pi^2}\, s^2\, e^{-4s^2/\pi}, \tag{8.2}$$

the spacing law of eigenvalues of large random Hermitian matrices, whose s^2 prefactor is the repulsion.

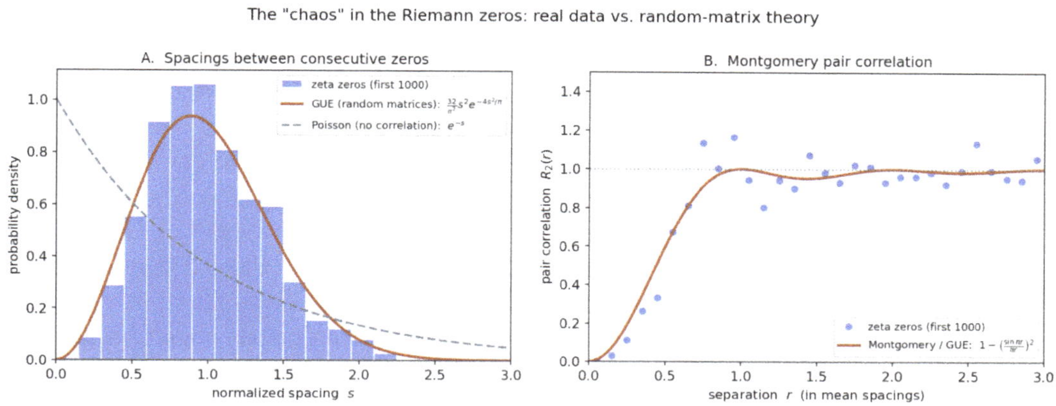

Figure 8.1: The zeros of ζ (first thousand, unfolded) against random-matrix theory. **A:** the nearest-neighbor spacing distribution. The data (bars) follow the GUE surmise (red), with its characteristic repulsion at small s, and look nothing like the uncorrelated Poisson law e^{-s} (dashed). **B:** the pair correlation. The data (points) follow Montgomery's law $1 - (\sin \pi u/\pi u)^2$ (red), including the deep suppression at small separation—the same correlation hole seen in the GUE.

8.4 Pair correlation

The second feature is the two-point statistic. Montgomery, studying the spacing of pairs of zeros in 1973, found that their pair correlation function is

$$R_2(u) = 1 - \left(\frac{\sin \pi u}{\pi u}\right)^2, \tag{8.3}$$

the very function that governs eigenvalue pairs in the GUE—a coincidence Freeman Dyson recognized on the spot. Figure 8.1B shows the first thousand zeros tracking Eq. (8.3): the correlation is strongly suppressed for separations below one mean spacing (zeros do not bunch), rises to 1, and oscillates gently around it. Montgomery proved the agreement for a restricted class of test functions; Odlyzko's high-precision computations, reaching enormous heights on the critical line, made the match overwhelming numerically; and Rudnick and Sarnak extended the agreement to higher correlations and to general L-functions, including the Dirichlet L-functions of the wheel [10, 18, 19].

8.5 Quantum chaos and the Hilbert–Pólya idea

It is worth stating the question this section answers in plain words before the names arrive. The zeros, we have just seen, carry the statistics of a spectrum—the energy levels of *something*. The natural question is: the spectrum of what? Three answers have been offered below, each attached to a name, and none has been delivered. What follows is the catalogue of those answers and the honest report that the object they reach for is still missing.

These two laws—level repulsion and the sine-kernel pair correlation—are the signature statistics of a particular kind of physical system. The Bohigas–Giannoni–Schmit conjecture holds that the quantized energy levels of a classically chaotic system with broken time-reversal symmetry obey GUE statistics [20]. The zeros, statistically, look exactly like such a spectrum. That resemblance revives a very old guess, attributed to Hilbert and Pólya: that the numbers γ_n are the eigenvalues of some self-adjoint operator H, so that $\frac{1}{2}+i\gamma_n$ are forced onto the critical line by the reality of a quantum spectrum. Berry and Keating proposed that the underlying classical dynamics might be the simple Hamiltonian $H \sim xp$; Connes gave a spectral realization on the *adele class space*—a single space that records every way of measuring the size of a rational number at once, the ordinary absolute value together with one scale for each prime—in which the zeros appear as an absorption spectrum [21, 22]. Universality ties this back to the wheel: the seven L-functions are not statistical outliers but members of the same ensemble, so each carries the same "spectrum" of zeros.

8.6 What this is and is not

What is established is the *statistics.* The local correlations of the zeros match those of the GUE; this is supported by theorems for restricted test functions and by numerics of extraordinary precision. It is genuine mathematics and it is not in doubt.

What is not established is the *operator.* There is no known quantum system—no Hamiltonian, no chaotic billiard, no dynamics—whose energy levels are the Riemann zeros. The Hilbert–Pólya statement is a conjecture supported by the statistical resemblance, not a construction; the proposals of Berry–Keating and Connes are programs, not realizations. "The zeros as a spectrum" is therefore the most physical-looking face of the subject and, at the same time, an analogy: the zeros behave as though they were a spectrum, and no one has exhibited the system for which they are one. The Riemann Hypothesis, which a genuine self-adjoint operator would settle at a stroke, remains open—for ζ and for every L-function on the wheel. This chapter shows that the zeros keep the statistics of a chaotic spectrum; it does not produce the chaos.

Reproducing the figure

```
import numpy as np
from mpmath import zetazero                    # generate the zeros (no data file)
gamma = np.sort(np.array([float(zetazero(k).imag) for k in range(1, 1001)]))

# unfold by the Riemann-von Mangoldt counting function -> unit mean spacing
def N(T): return T/(2*np.pi)*np.log(T/(2*np.pi)) - T/(2*np.pi) + 7/8
u = N(gamma)
s = np.diff(u)
print(s.mean(), s.std())          # 1.000, 0.38
print(np.mean(s < 0.1), np.mean(s < 0.5))   # 0.000, 0.073   (Poisson: 0.39)

# Panel A: histogram of s vs GUE surmise and Poisson
#   p_GUE(s) = (32/pi**2) s**2 exp(-4 s**2/pi);  p_Poisson(s) = exp(-s)
```

```
# Panel B: pair correlation of u vs  1 - (sin(pi r)/(pi r))**2
```

Chapter 9

The Supersymmetric Primon Gas

This is an expository *capstone to the earlier chapters on the mod-*210 *wheel. It assembles a single object from pieces that are all classical. A bosonic gas with one mode per prime, twisted by a Dirichlet character χ, has partition function the L-function $L(s,\chi)$ — the same object whose zeros the earlier chapters drew as a quasicrystal. Its fermionic superpartner, in which Pauli exclusion restricts the states to the squarefree integers and the fermion number is graded by $(-1)^F$, has partition function $1/L(s,\chi) = \sum_n \chi(n)\mu(n)\, n^{-s}$. The two are joined by supersymmetry, and the boson–fermion cancellation is exactly the (twisted) Möbius inversion identity; the grading $(-1)^F$* is *the twisted Möbius function $\chi\mu$. On the wheel this unifies the bosonic gas, the seven real L-functions, and the Möbius function as one supersymmetric system, and the conductor of each character silences its primes in both the diffraction picture and the Möbius picture — the same lattice indexing both. None of this is new: the construction is the supersymmetric primon gas of Spector and Julia [24, 23], with the arithmetic-gas twisting of Bakas and Bowick [25] and its recent L-function generalization [26]. The output is the Möbius function; the spectrum is the primes; the Riemann zeros do not appear.*

Before any of the physics, here is the whole chapter in one plain sentence. The wheel's L-function is a sum over the integers; its reciprocal is the same sum with Möbius signs attached; the two multiply to 1, and that statement *is* Möbius inversion. Everything that follows—the gas, the bosons and fermions, the supersymmetry, the Witten index—is a physical costume worn by that one identity. The costume is real physics and it is the reason the construction is beautiful, but it is a costume: the content is the inversion, an object the wheel already owns from the Möbius function. A reader who wants only the mathematics can take $Z_B = L$, the signed $Z_F = 1/L$, and their product 1, and has lost nothing. The rest of the chapter dresses those three facts and ties the dressing back to the conductor lattice.

9.1 Scope

This chapter states no theorem and claims nothing new. It is a synthesis: it puts together the bosonic primon gas (Julia [23]), its supersymmetric completion and the resulting Möbius function (Spector [24]), the arithmetic-gas twisting machinery (Bakas and Bowick [25]), and the by-now-standard fact that a Dirichlet L-function is the partition function of a character-twisted primon gas [25] (see also [26]). The only thing local to this chapter is that the assembly is carried out on the mod-210 wheel and tied to the earlier chapters on the prime race, the diffraction picture, and the conductor lattice. A reader who wants the physics in full should see the review of Schumayer and Hutchinson [27].

9.2 The bosonic primon gas

Take a bosonic Fock space with one mode per prime p. A basis state is a finite multiset of primes, equivalently a positive integer $n = \prod_p p^{a_p}$, and we assign it energy $E_n = \log n$, so that $e^{-\beta E_n} = n^{-s}$ with $s = \beta$. The partition function is then

$$Z_B(s) = \sum_{n\geq 1} n^{-s} = \prod_p \left(1 - p^{-s}\right)^{-1} = \zeta(s), \tag{9.1}$$

the Euler product being unique factorization read as "each mode independently occupied any number of times." This is Julia's free Riemann gas [23]: the bosonic Fock space over the primes, with the integers as its occupation-number states.

Twisting each mode by a Dirichlet character—giving the primon of prime p a charge $\chi(p)$—replaces ζ by the L-function,

$$Z_B(s, \chi) = \prod_p \left(1 - \chi(p)p^{-s}\right)^{-1} = \sum_{n\geq 1} \chi(n)\, n^{-s} = L(s, \chi), \tag{9.2}$$

valid because χ is completely multiplicative. The L-functions of the diffraction and conductor-lattice chapters are therefore nothing other than the partition functions of character-twisted primon gases.

9.3 The fermionic superpartner

Now give each prime mode a fermionic partner subject to the Pauli exclusion principle: each prime may occupy its fermionic mode at most once. The accessible states are then the *squarefree* integers, and the partition function is

$$Z_F(s, \chi) = \prod_p \left(1 + \chi(p)p^{-s}\right) = \prod_p \frac{1 - \chi(p)^2 p^{-2s}}{1 - \chi(p)p^{-s}} = \frac{L(s, \chi)}{L(2s, \chi^2)}. \tag{9.3}$$

The construction that matters for the marriage is the *signed* fermionic sum, in which each state is weighted by the fermion parity $(-1)^F$. Since a squarefree $n = p_1 \cdots p_k$ carries k fermions, its weight is $(-1)^k \chi(n) = \chi(n)\mu(n)$, and

$$\prod_p \left(1 - \chi(p)p^{-s}\right) = \sum_{n\geq 1} \chi(n)\mu(n)\, n^{-s} = \frac{1}{L(s, \chi)}. \tag{9.4}$$

So the signed fermionic gas is the reciprocal L-function, and its Dirichlet coefficients are the *twisted Möbius function* $\chi(n)\mu(n)$, supported on the squarefree integers and signed by the parity of the number of prime factors. This is Spector's observation [24], twisted by a character.

9.4 The marriage: supersymmetry

The grading $(-1)^F$ sounds like physics and means something elementary: count each state with a plus sign if it is built from an even number of distinct primes and a minus sign if from an odd number—exactly the sign the Möbius function already assigns. With that bookkeeping the boson and fermion partners cancel in pairs, and only the empty state survives.

Join the two gases as superpartners and grade the combined trace by $(-1)^F$. Because the bosonic states multiply the signed fermionic states, the graded partition function factorizes,

$$\mathrm{Tr}\big[(-1)^F e^{-\beta H}\big] = Z_B(s,\chi)\cdot\sum_n \chi(n)\mu(n)\,n^{-s} = L(s,\chi)\cdot\frac{1}{L(s,\chi)} = 1. \tag{9.5}$$

Every state above the vacuum cancels against a partner of opposite statistics; only the vacuum ($n = 1$) survives, giving Witten index 1. Read coefficient by coefficient, Eq. (9.5) is the (twisted) Möbius inversion identity

$$\sum_{d|n} \chi(d)\,\chi(n/d)\mu(n/d) = \chi(n)\sum_{d|n}\mu(d) = [\,n = 1\,]. \tag{9.6}$$

The boson–fermion cancellation *is* Möbius inversion, and the grading $(-1)^F$ *is* the twisted Möbius function. Table 9.1 records the three identities, evaluated numerically at $s = 2$ for the trivial character and for two of the wheel's real characters; the partition functions and their reciprocals multiply to 1 to machine precision, and the per-integer cancellation is exact.

Table 9.1: Numerical check at $s = 2$. The bosonic partition function (Euler product) equals $L(s,\chi)$; the signed fermionic sum $\sum_n \chi(n)\mu(n)n^{-s}$ equals $1/L(s,\chi)$; their product is 1. The trivial character recovers $\zeta(2) = \pi^2/6$ and $1/\zeta(2) = 6/\pi^2$.

character	$L(2,\chi)$	$\sum_n \chi(n)\mu(n)/n^2$	$1/L(2,\chi)$	product
trivial (ζ)	1.644934	0.607927	0.607927	1.000000
$\chi_{5.4}$ (D=+5)	0.706211	1.416007	1.416007	1.000000
$\chi_{7.6}$ (D=−7)	1.151925	0.868112	0.868112	1.000000

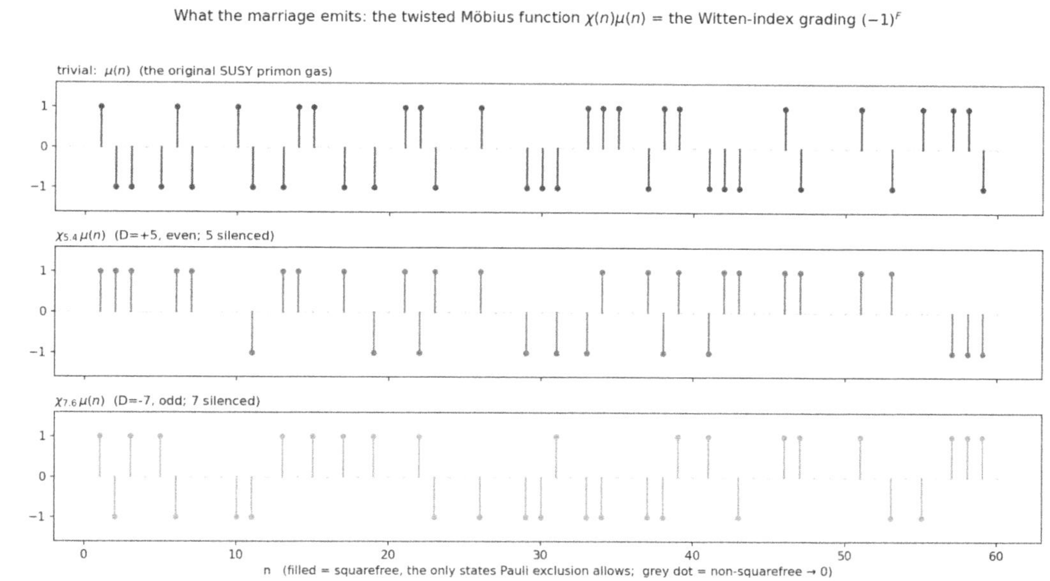

Figure 9.1: What the marriage emits: the twisted Möbius function $\chi(n)\mu(n)$, which is the fermion-parity grading $(-1)^F$ and the Dirichlet coefficient of $1/L(s,\chi)$. It is nonzero only on the squarefree integers — the only states Pauli exclusion allows — and signed by χ and the parity of the number of prime factors. Top: the untwisted Möbius function $\mu(n)$ (Spector's original gas). Middle and bottom: the twists by the conductor-5 and conductor-7 characters; each silences the multiples of its own conductor prime.

9.5 On the wheel

Each of the seven real characters modulo 210 thus carries its own supersymmetric primon gas: a bosonic gas whose partition function is its L-function, and a fermionic superpartner whose signed partition function is the reciprocal, with coefficients the twisted Möbius function. The seven L-functions of the conductor-lattice chapter are the bosonic halves; the seven twisted Möbius functions are the fermionic halves.

The conductor reappears in the same role it played in the diffraction picture. There, a prime dividing the conductor switched off its diffraction peak because $\chi(p) = 0$; here, the same vanishing removes that prime from the twisted Möbius function (Fig. 9.1: $\chi_{5.4}\mu$ has no support on multiples of 5, $\chi_{7.6}\mu$ none on multiples of 7). The conductor lattice indexes both pictures at once: it says which primes are silenced, whether one is looking at the spectral side (the zeros' diffraction) or the arithmetic side (the Möbius function). The quasicrystal and the supersymmetric gas are the same wheel seen from two directions.

It is worth saying which composites each side actually sees, since the two are easy to confuse. The diffraction peaks sit at prime powers—the support of the von Mangoldt function Λ—so that picture registers the prime *squares* $121 = 11^2$ and $169 = 13^2$ as faint echoes near $2\log 11$ and $2\log 13$, and is blind to the squarefree composites. The Möbius function μ has exactly the complementary support, the *squarefree* integers, so the gas is blind to those two squares and instead registers the three squarefree composites $143 = 11 \cdot 13$, $187 = 11 \cdot 17$, $209 = 11 \cdot 19$ (each with $\mu = +1$). The wheel's five composite spokes thus divide cleanly between the pictures—$121, 169$ seen only in the diffraction, $143, 187, 209$ only in the gas—while the primes are untouched by either subtlety, being the full-strength peaks of the one and the $\mu = -1$ generators of the other.

9.6 What this is and is not

The construction is classical throughout. The bosonic gas is Julia's [23]; the supersymmetric completion and the Möbius function as Witten index are Spector's [24]; the twisting and boson–fermion equivalence are Bakas and Bowick's [25]; the identification of a Dirichlet L-function with a character-twisted primon partition function is theirs as well [25], recently generalized [26]. This chapter only assembles them on one modulus and ties them to the earlier chapters. It is a reformulation, not a result: we supplied the primon gas (known) and the L-functions (known), and out came the twisted Möbius function (the known corollary). Nothing emerged that was not placed.

In particular it does not approach the Riemann zeros. The spectrum of the gas, bosonic or supersymmetric, is the set of energies $\{\log n\}$ — the primes and their products. The partition functions are $L(s, \chi)$ and $1/L(s, \chi)$; the nontrivial zeros live in the analytic continuation of those functions, not among the energy levels of the Hamiltonian. So the marriage manufactures the Möbius function, which is itself built from prime factorization, and unifies the bosonic and fermionic descriptions of the wheel beautifully — and it remains entirely on the elementary side of the wall. Producing the zeros as an actual spectrum is the separate, unsolved problem of Hilbert and Pólya, untouched here.

Reproducing the check

```
import numpy as np
N = 2_000_000; s = 2.0
mu = np.ones(N+1, dtype=np.int64); mu[0]=0          # linear/Eratosthenes Mobius sieve
sieve = np.ones(N+1, bool); sieve[:2]=False
for i in range(2, int(N**0.5)+1):
    if sieve[i]: sieve[i*i::i]=False
for p in np.nonzero(sieve)[0]:
    mu[p::p] *= -1; mu[p*p::p*p] = 0

def chi(conductor, plus):                  # real character by residue mod conductor
    idx = np.arange(N+1) % conductor
    tab = np.array([0 if np.gcd(r,conductor)!=1 else (1 if r in plus else -1)
                    for r in range(conductor)])
    return tab[idx]

inv = np.zeros(N+1); inv[1:] = np.arange(1,N+1)**(-s)
```

```
for name, c, plus in [("zeta",1,set()), ("chi_5.4",5,{1,4}), ("chi_7.6",7,{1,2,4})]:
    x = np.ones(N+1) if c==1 else chi(c, plus)
    L      = (x[1:]*inv[1:]).sum()                    # bosonic: L(s,chi)
    invL   = (x[1:]*mu[1:]*inv[1:]).sum()    # signed fermionic: sum chi*mu/n^s
    print(name, L, invL, L*invL)                       # product -> 1
```

Part IV

Climbing and counting

Chapter 10

Climbing the Primorials

The opening chapter fixed one wheel, $M = 210$. This chapter asks what happens to its 48 spokes as one walks along the number line, and as one climbs to larger primorials. Two classical facts answer it. An exact conservation law holds in every period: the unit, the primes, and the coprime composites always sum to $\varphi(M)$, so that as the primes thin out (by the prime number theorem in progressions) the composites fill the spokes they vacate. And the wheel's density itself thins at the rate fixed by Mertens' third theorem, $\prod_{p \le X}(1 - 1/p) \sim e^{-\gamma}/\log X$, which a short identity ties to the Hardy–Littlewood twin constant of the twins chapter. Everything is classical—Mertens (1874), the prime number theorem in progressions—and none of it touches the Riemann Hypothesis. This chapter is expository.

10.1 Scope

This chapter is expository and classical [28, 2, 29]. It extends the opening chapter, which described the single wheel $M = 210$, in two directions—along the integers (the per-period census) and up the primorials (the density)—and it connects forward to the twin constant of the twins chapter. It states no new result. As Sec. 10.5 records, the content here is exactly the kind that depends only on the prime number theorem and Mertens' theorems, with no bearing on the Riemann Hypothesis.

10.2 The census: a conservation law

Fix a primorial M and walk through the integers one period at a time, the kth period being the block $(kM, (k+1)M]$. Each integer in the block that is coprime to M is one of exactly three things: the unit 1 (only in the first block), a prime, or a composite coprime to M. Writing $U(k)$, $P(k)$, $S(k)$ for their counts, this trichotomy is a conservation law,

$$U(k) + P(k) + S(k) = \varphi(M) \qquad \text{for every } k, \tag{10.1}$$

exact in every period because each of the $\varphi(M)$ coprime residue classes is represented once per block. For $M = 210$ the total is always 48 (verified across all 100 periods up to 21000, and the same is true at $M = 2310$). Figure 10.1 shows the stacked counts: every bar reaches 48.

What changes from block to block is the split. By the prime number theorem in arithmetic progressions, the primes in a block near height kM number

$$P(k) \sim \frac{M}{\log\big((k+\frac{1}{2})M\big)}, \tag{10.2}$$

which decreases as k grows; and since the total is pinned at $\varphi(M)$, the coprime composites $S(k) = \varphi(M) - P(k) - U(k)$ must increase to compensate, rising toward $\varphi(M)$. The primes that vacate the spokes are replaced by composites; the wheel fills in. The smooth crossover $P = S$ occurs where $M/\log(kM) = \varphi/2 = 24$, near $k \approx 30$, though single-period counts fluctuate around this trend.

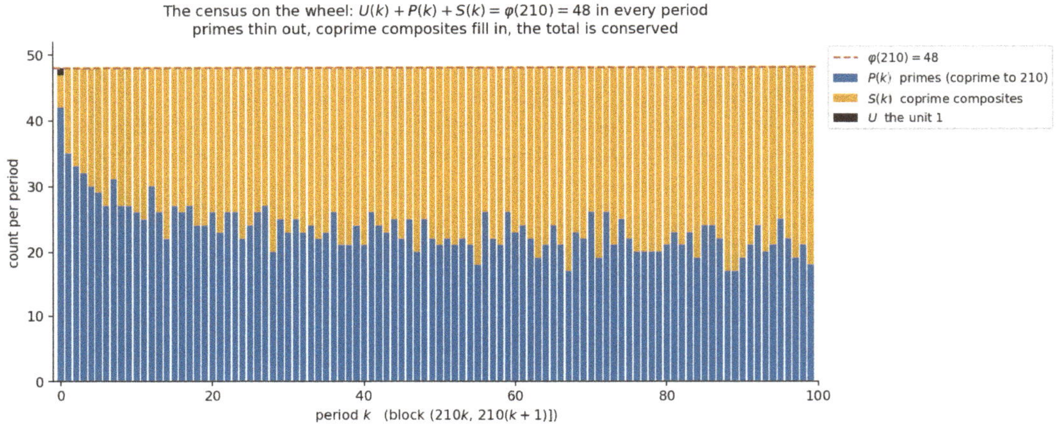

Figure 10.1: The census at $M = 210$. In each period the unit, the primes $P(k)$ (blue), and the coprime composites $S(k)$ (orange) sum to $\varphi(210) = 48$ (red line): the conservation law (10.1). As one moves out, $P(k)$ thins by the prime number theorem in progressions and $S(k)$ fills the vacated spokes; the total never moves.

10.3 Mertens' theorems and the thinning

The density of the wheel itself—the fraction of integers that are coprime to M—thins as M grows, and the rate is classical. Mertens' third theorem states

$$\prod_{p \le X}\left(1 - \frac{1}{p}\right) \sim \frac{e^{-\gamma}}{\log X}, \tag{10.3}$$

with γ the Euler–Mascheroni constant. Equivalently the product times $\log X$ approaches the constant $e^{-\gamma} = 0.5615\ldots$; Fig. 10.2 shows it doing so, approaching from below and settling by $X \sim 10^3$. Since the kth primorial M_k has $\log M_k \sim p_k$

(Chebyshev), the wheel's density obeys

$$\frac{\varphi(M_k)}{M_k} = \prod_{p \le p_k} \left(1 - \frac{1}{p}\right) \sim \frac{e^{-\gamma}}{\log p_k}, \tag{10.4}$$

the slow decay already pictured for the first few primorials in the opening chapter. (Mertens' second theorem, $\sum_{p \le X} 1/p = \log\log X + B$ with $B = 0.2615\ldots$, is the additive companion and governs the finer corrections.)

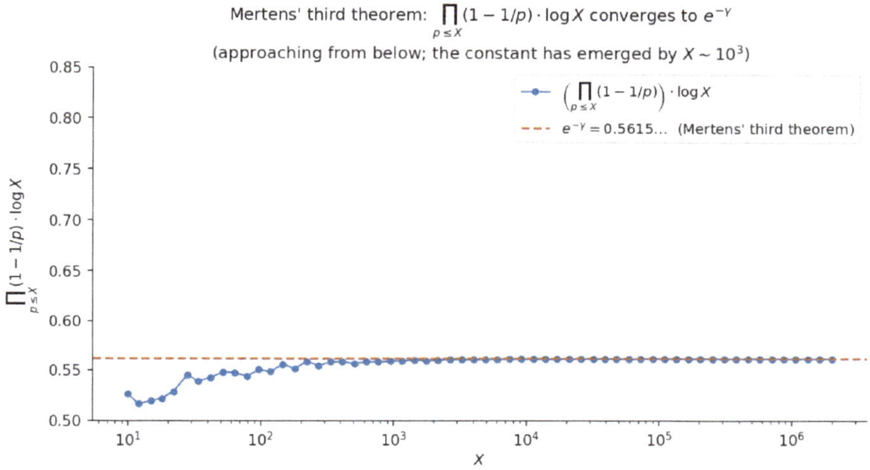

Figure 10.2: Mertens' third theorem (10.3) in the form $\prod_{p \le X}(1 - 1/p) \cdot \log X \to e^{-\gamma}$. The product approaches the constant from below and has effectively reached it by $X \sim 10^3$.

10.4 A bridge to the twin constant

The same density connects to the twins chapter through one short identity. For each prime $p \ge 3$,

$$\left(1 - \frac{1}{(p-1)^2}\right)\left(1 - \frac{1}{p}\right) = \frac{p-2}{p-1}, \tag{10.5}$$

a one-line algebraic fact. Taking the product over $3 \le p \le X$ turns it into

$$\prod_{3 \le p \le X} \frac{p-2}{p-1} = C_2 \prod_{3 \le p \le X} \left(1 - \frac{1}{p}\right)(1 + o(1)), \qquad C_2 = \prod_{p \ge 3}\left(1 - \frac{1}{(p-1)^2}\right) = 0.6601\ldots, \tag{10.6}$$

where C_2 is the Hardy–Littlewood twin-prime constant (the two sides agree to ratio 1.0000 already by $X \approx 2 \times 10^4$). Combining Eq. (10.6) with Mertens' third theorem gives the twin-admissible density of the wheel, $\prod_{3 \le p \le X}(p-2)/(p-1) \sim 2C_2 e^{-\gamma}/\log X$. This product is exactly the singular-series factor that sets the rate of twin primes in the twins chapter: the density that thins the wheel here is the same one that, refined by C_2, predicts how often the twin slots are filled.

10.5 What this is and is not

All of this is classical and self-contained. The conservation law (10.1) is elementary; the decay of $P(k)$ is the prime number theorem in progressions; the density rate is Mertens (1874); the twin-constant identity is a one-line computation. None of it depends on the Riemann Hypothesis or on any unproven conjecture—it uses only the prime number theorem, Mertens' theorems, and the classical zero-free region.

The census and the density developed here are simply facts about the wheel—true whether or not anything is built on top of them, and dependent only on the prime number theorem and Mertens' theorems. That is why they belong in the tour.

Reproducing the figures

```
import numpy as np
from math import gcd

M = 210; nper = 100; N = M*nper
s = np.ones(N+1, bool); s[:2] = False
for i in range(2, int(N**0.5)+1):
    if s[i]: s[i*i::i] = False
phi = sum(1 for r in range(M) if gcd(r, M) == 1)        # 48

for k in range(nper):                                    # conservation law
    U = P = S = 0
    for n in range(k*M+1, (k+1)*M+1):
        if gcd(n, M) != 1: continue
        if n == 1: U += 1
        elif s[n]: P += 1
        else: S += 1
    assert U + P + S == phi                              # holds every period

# Mertens' third:  prod_{p<=X}(1-1/p) * log X  ->  e^{-gamma}
import numpy as np
pr = np.nonzero(s)[0]
prod = np.cumprod(1 - 1/pr.astype(float))
# prod[-1]*log(N) approaches exp(-0.5772156649) = 0.5615

# Lemma:  prod_{3<=p<=X}(p-2)/(p-1) = C2 * prod(1-1/p) * (1+o(1))
C2 = 0.6601618158468695
odd = pr[(pr >= 3) & (pr <= N)]
lhs = np.prod((odd-2)/(odd-1)); rhs = C2*np.prod(1 - 1/odd.astype(float))
print(lhs/rhs)                                           # ~ 1.0000
```

Chapter 11

The Repeating Decimals

Every residue r coprime to 210 gives a fraction $r/210$ whose decimal expansion repeats with a single digit of delay and a period of six. This short, self-contained chapter explains why: $210 = 10 \times 21$, the one delayed digit comes from the factor $10 = 2 \cdot 5$ shared with the base, and the period six is the multiplicative order of 10 modulo 21. The 48 coprime residues therefore carry only twelve distinct six-digit blocks, and those twelve fall into two *cyclic families—six rotations each of a single "cyclic number," both of which are the famous $1/7 = 0.\overline{142857}$ in disguise. We show that the family of a residue is nothing more elaborate than its value modulo 3; that the wheel's mirror $r \leftrightarrow 210 - r$ acts on the decimals as the digit-wise nines-complement while flipping the family; and that walking up the periods and walking across the residues are two perpendicular motions—one additive, one multiplicative. Everything here is classical (the theory of repeating decimals and cyclic numbers), and everything is a base-10 lens on the genuinely base-independent object underneath: the multiplicative group $(\mathbb{Z}/210\mathbb{Z})^\times$. No new theorem is claimed. This chapter is the decimal counterpart to the opening chapter and the two-mirrors chapter, and it points, for the part that is actually deep, to the chapters on how the primes distribute across these classes.*

11.1 Scope

This is an expository chapter. Its purpose is to make one piece of the mod-210 wheel vivid in base 10: the repeating decimals attached to the coprime residues. The opening chapter introduces the 48 units of $(\mathbb{Z}/210\mathbb{Z})^\times$ and the reflection $r \leftrightarrow 210 - r$; the two-mirrors chapter reads that reflection as the analytic functional equation. Here we add nothing to the arithmetic. We simply photograph the same residues through the lens of their decimal expansions and observe that the photograph is unusually orderly—twelve blocks, two families, a clean mirror law—and then we are honest about what that order is and is not. The order is real and classical; the decimals themselves are an artifact of writing numbers in base ten. The object they shadow—the multiplicative group and its behaviour on the primes—is what the rest of the book is about.

11.2 Why one digit of delay and a period of six

Write the modulus as

$$210 = 2 \cdot 3 \cdot 5 \cdot 7 = \underbrace{(2 \cdot 5)}_{10} \times \underbrace{(3 \cdot 7)}_{21} = 10 \times 21. \tag{11.1}$$

The factor 10 is exactly the part of the modulus that shares its primes with the base. For any reduced fraction the decimal expansion consists of a non-repeating *pre-period* followed by a repeating *period*, and the two lengths are governed separately: the pre-period length is $\max(v_2(210), v_5(210)) = \max(1, 1) = 1$, and the period length is the multiplicative order of 10 modulo the part coprime to the base, namely 21—the dependence of the period on this order being Gauss's result [30]. Computing that order,

$$10^1 \equiv 10, \quad 10^2 \equiv 16, \quad 10^3 \equiv 13, \quad 10^4 \equiv 4, \quad 10^5 \equiv 19, \quad 10^6 \equiv 1 \pmod{21}, \tag{11.2}$$

so $\text{ord}_{21}(10) = 6$. Hence every $r/210$ with $\gcd(r, 210) = 1$ looks like

$$\frac{r}{210} = 0.\,d\,\overline{b_1 b_2 b_3 b_4 b_5 b_6}, \tag{11.3}$$

one delayed digit d and then a six-digit block on infinite repeat. For example $\frac{11}{210} = 0.0\overline{523809}$.

11.3 Twelve blocks, two cyclic families

Because the repeating part is controlled entirely by the order of 10 modulo 21, the six-digit block depends only on $r \bmod 21$. There are $\varphi(21) = \varphi(3)\varphi(7) = 2 \cdot 6 = 12$ residues coprime to 21, so there are exactly *twelve* distinct blocks. Each is shared by $48/12 = 4$ of the residues coprime to 210 (those reducing to the same class mod 21). Table 11.1 lists all twelve with their residues.

The twelve residues mod 21 are organized by the cyclic subgroup $\langle 10 \rangle$. Since $\text{ord}_{21}(10) = 6$, that subgroup has order 6 and index 2 in the group of order 12; the group therefore splits into *two* cosets of six elements each. Each coset is one orbit under multiplication by 10—and multiplication by 10 is precisely a one-step cyclic rotation of the decimal block. So each family is a single "cyclic number" read from six different starting points (Fig. 11.1):

$$\textbf{Family A}: \quad 047619 \to 476190 \to 761904 \to 619047 \to 190476 \to 904761, \tag{11.4}$$

$$\textbf{Family B}: \quad 095238 \to 952380 \to 523809 \to 238095 \to 380952 \to 809523. \tag{11.5}$$

Both are the most famous cyclic number of all in disguise. Since $21 = 3 \cdot 7$,

$$\frac{1}{21} = \frac{1}{3} \cdot \frac{1}{7} = 0.\overline{047619}, \qquad 047619 = \frac{142857}{3}, \qquad 095238 = 2 \times 047619, \tag{11.6}$$

where 142857 is the cyclic number of $1/7 = 0.\overline{142857}$. The whole decimal life of the wheel is one sixth of the seven-cycle, taken once (Family A) and twice (Family B).

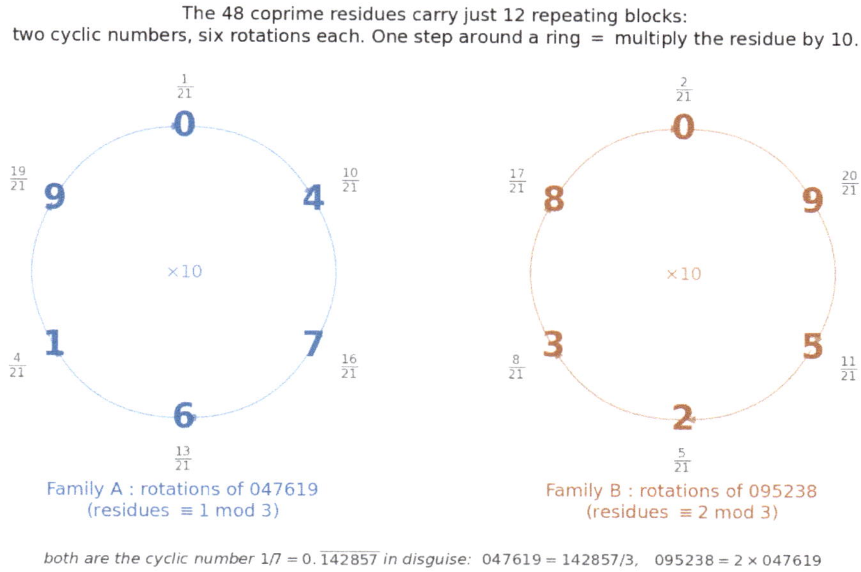

Figure 11.1: The 48 coprime residues carry only twelve repeating blocks, and those twelve are two cyclic numbers read from six starting points. Each digit on a ring is a rotation; one step clockwise multiplies the residue (shown as $r/21$) by 10. Family A is one copy of the $1/7$ cycle, Family B is two copies. The split is exactly residue mod 3 (Sec. 11.4).

Table 11.1: The twelve repeating blocks of the mod-210 wheel. Each block is the six-digit period of $r/210$ for the four listed residues; "family" is the coset of $\langle 10 \rangle$, which coincides with residue mod 3; "mirror block" is the period of $(210 - r)/210$ and is the digit-wise nines-complement (Sec. 11.5).

Block	Family	Residues mod 210	Mirror block
047619	A	1, 43, 127, 169	952380
190476	A	67, 109, 151, 193	809523
476190	A	31, 73, 157, 199	523809
619047	A	13, 97, 139, 181	380952
761904	A	37, 79, 121, 163	238095
904761	A	19, 61, 103, 187	095238
095238	B	23, 107, 149, 191	904761
238095	B	47, 89, 131, 173	761904
380952	B	29, 71, 113, 197	619047
523809	B	11, 53, 137, 179	476190
809523	B	17, 59, 101, 143	190476
952380	B	41, 83, 167, 209	047619

11.4 The family is the mod-3 clock

The two families look like a fact about decimals, but they are a fact about residues mod 3. The family of r is its coset of $\langle 10 \rangle$ in $(\mathbb{Z}/21\mathbb{Z})^{\times}$. Multiplication by 10 moves within a coset, and

$$10 \equiv 1 \pmod{3}, \tag{11.7}$$

so multiplying by 10 leaves r mod 3 unchanged. The two cosets are therefore separated by exactly one invariant—the value modulo 3:

$$\textbf{Family A} \iff r \equiv 1 \pmod{3}, \qquad \textbf{Family B} \iff r \equiv 2 \pmod{3}. \tag{11.8}$$

This is the same mod-3 clock that runs through *Twin Primes on the Mod-210 Wheel*. There, the lower member of every twin pair is $\equiv 2 \pmod{3}$ and the upper is $\equiv 1 \pmod{3}$ (the integer between them carries the factor of 3), so every twin corridor reads "$B \to A$." The decimal families and the twin admissibility are two faces of the single elementary fact that the units mod 3 are $\{1, 2\}$.

11.5 The mirror is the nines-complement

The wheel's reflection $r \leftrightarrow 210 - r$ has an immediate decimal meaning. Since

$$\frac{r}{210} + \frac{210 - r}{210} = 1 = 0.\overline{9}, \tag{11.9}$$

the two repeating blocks must add, digit by digit, to all nines: their sum is 999999 (Fig. 11.2). The mirror also flips the family. Negation by -1 sends a coset of $\langle 10 \rangle$ to the other coset, because $-1 \equiv 2 \pmod{3}$ lies in Family B; so reflection always sends $A \leftrightarrow B$. In Table 11.1 every mirror block sits in the opposite family from its partner, and the two blocks are nines-complements. This is the decimal shadow of exactly the reflection that *The Two Mirrors* matches to the $s \leftrightarrow 1 - s$ symmetry of the completed zeta function.

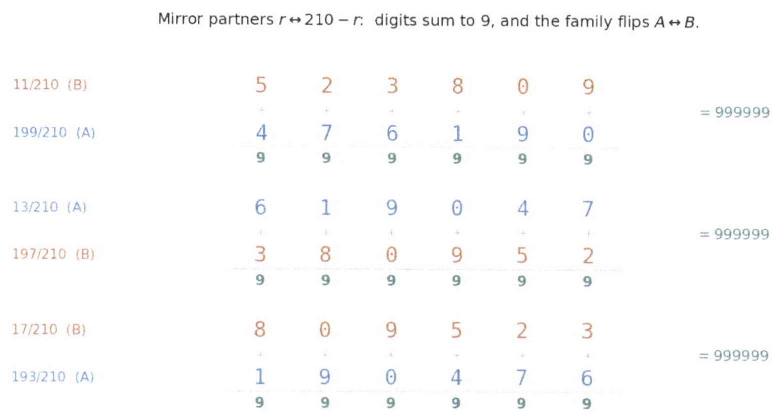

Figure 11.2: Mirror partners r and $210 - r$. Their repeating blocks are digit-wise nines-complements (each column sums to 9, the whole block to 999999), and the family flips $A \leftrightarrow B$.

11.6 Two perpendicular axes

A number coprime to 210 has two coordinates, and the decimal displays them on perpendicular axes (Fig. 11.3). Writing

$$n = 210\,k + r, \qquad \frac{n}{210} = k + \frac{r}{210}, \tag{11.10}$$

the integer part is the *period* k and the fractional part is fixed by the *residue* r. These move independently:

- **Period axis (additive).** Increase k with r fixed: the integer in front climbs $0, 1, 2, \ldots$ while the repeating block stays frozen. This is place value; it is the elementary coordinate.
- **Residue axis (multiplicative).** Replace r by $10r \bmod 21$ within a period: the integer stays put while the block rotates one step. This is motion inside the group $(\mathbb{Z}/210\mathbb{Z})^{\times}$; it is the cyclic structure of Sec. 3.

Walking up the wheel's periods advances the front and holds the tail; walking around its residues rotates the tail and holds the front. The two are genuinely orthogonal: one is counting, the other is the multiplicative group at work.

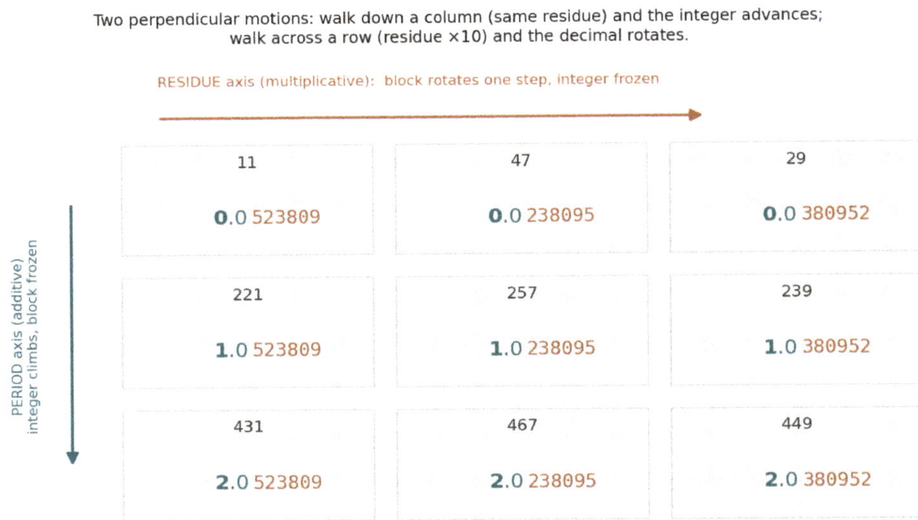

Figure 11.3: The two perpendicular motions. Down a column the residue is fixed, the integer climbs $0, 1, 2$ and the block is frozen. Across a row the residue is multiplied by 10, the integer is frozen and the block rotates ($523809 \to 238095 \to 380952$). Period is the additive axis; residue is the multiplicative axis.

11.7 What this is, and what it is not

What it is. A correct, classical description of the repeating decimals of $r/210$: one delayed digit, period six, twelve blocks, two cyclic families that are the $1/7$ cycle scaled, a family invariant equal to residue mod 3, and a mirror law that is the digit-wise nines-complement with a family flip. All of it follows from the theory of decimal expansions and the order of 10 modulo 21—material that traces back to Gauss, who worked out the periods of fractions and their dependence on the order of the base in Section VI of the *Disquisitiones Arithmeticae* [30], and which has been a staple of elementary number theory and the study of cyclic numbers ever since [1]. It is a clean and, I think, beautiful way to organize the residues—and through them the primes and the coprime composites—on the wheel.

What it is not. It is not base-independent, and it is not new mathematics. The very existence of a six-digit block, and the particular digits in it, are artifacts of writing numbers in base 10: change the base b and the period becomes the order of b modulo the part of 210 coprime to b, the blocks regroup, and the pretty 142857 connection dissolves. In base 60—the base of a clock—the 3 is absorbed into the terminating part, leaving the 7 alone to drive a period-*three* repeat, and the families stop being residue mod 3 and become the quadratic-residue character mod 7: the same forty-eight numbers re-sorted by a different prime entirely (Fig. 11.4). The object that does *not* depend on the base—the thing the decimals are a shadow of—is the multiplicative group $(\mathbb{Z}/210\mathbb{Z})^\times \cong \mathbb{Z}/2 \times \mathbb{Z}/4 \times \mathbb{Z}/6$ together with its mod-3 quotient and its reflection. Those are real and they recur throughout this book. Nothing in this chapter is a theorem that was not already known, and nothing here bears on factorization or on the twin prime conjecture: the residue, the period, and the decimal are all functions of n, and a function of n cannot reveal more than n itself contains. The genuinely deep questions are not about the digits but about how the primes *distribute* across the twelve blocks and the 48 classes, and those belong to the chapters on equidistribution, the prime race, and the connection of the wheel to the zeros. This chapter is the lens; those chapters are the landscape.

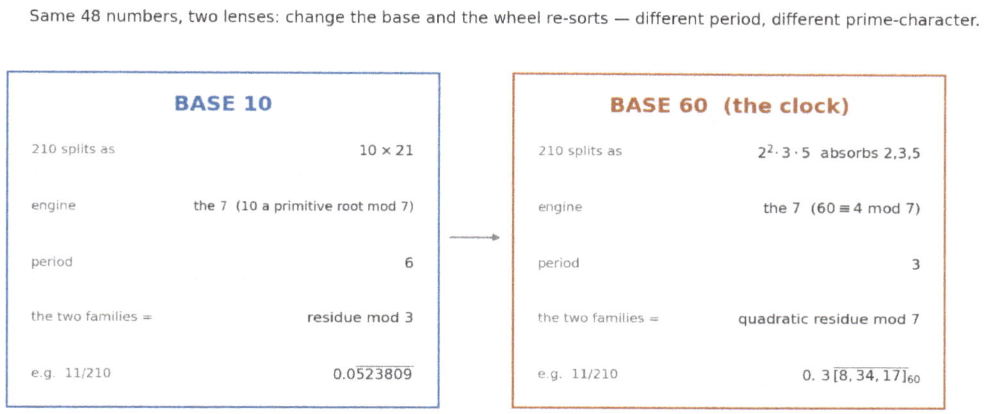

Figure 11.4: The same forty-eight numbers under two bases. In base 10 the repeating engine is the 7 (where 10 is a primitive root), the period is six, and the families are residue mod 3. In base 60—a clock's base—the 3 is absorbed into the terminating head, leaving the 7 alone to drive a period-three repeat, and the families become the quadratic-residue character mod 7. The period and the family-character are properties of the lens, not of the numbers; only the underlying group is invariant.

Reproducible computation

The blocks, families, mirror partners, and the order of 10 are reproduced by the following self-contained Python.

```
from math import gcd

def block(r):                      # six-digit repeating block of r/210
    x, s = r % 21, ""              # period is governed by the part coprime to 10, i.e. 21
    for _ in range(6):
        x *= 10; s += str(x // 21); x %= 21
    return s

units = [r for r in range(1, 210) if gcd(r, 210) == 1]    # 48 of them

groups = {}
```

```
for r in units:
    groups.setdefault(block(r), []).append(r)

assert len(groups) == 12 and all(len(v) == 4 for v in groups.values())

for b in sorted(groups):
    r0  = groups[b][0]
    fam = 'A' if r0 % 3 == 1 else 'B'          # family = residue mod 3
    mir = block(210 - r0)                      # mirror block
    assert int(b) + int(mir) == 999999         # nines-complement
    print(b, fam, sorted(groups[b]), mir)

# order of 10 modulo 21 is 6
o, x = 0, 1
while True:
    x = (x * 10) % 21; o += 1
    if x == 1: break
assert o == 6
```

Chapter 12

Twin Primes

This is the capstone of the tour, and it returns to the question the wheel was built around: twin primes. A pair $(p, p+2)$ can both be prime only if both residues are coprime to 210, and exactly 15 of the 48 spokes begin such an admissible slot. These 15 slots are perfectly mirror-symmetric under $r \mapsto 210 - r$. The primes, however, are not. Among the residues themselves, six of the seven mirror pairs hold two genuine twins, but one is broken: $(41, 43)$ is a twin while its mirror $(167, 169)$ is not, because $169 = 13^2$. A single square spoils one half of a symmetric pair. The Hardy–Littlewood heuristic predicts how the twins are distributed across these slots and how many there are—a prediction the data confirm to within half a percent at 10^7—but whether the twins are infinite at all is the twin prime conjecture, still open. This chapter is expository: a well-tested heuristic and an exact structural observation, resting on a question no one can yet answer.

12.1 Scope

This chapter is expository. The Hardy–Littlewood heuristic it describes is classical [31]; the slot structure and its mirror symmetry are elementary; the numerical agreement is reproduced in the appendix. It proves nothing, and it is careful in Sec. 12.5 about the one thing that matters most here: the infinitude of twin primes is conjectural. This chapter draws on the opening chapter for the wheel and on the two-mirrors chapter for the reflection, and it closes the tour.

12.2 The twin slots

A twin prime is a pair $(p, p+2)$ with both members prime. Past 7, each member must be coprime to 210, so a twin can occupy only those residues r for which both r and $r+2$ are coprime to 210. Counting these prime by prime,

$$\underbrace{1}_{p=2} \cdot \underbrace{1}_{p=3} \cdot \underbrace{3}_{p=5} \cdot \underbrace{5}_{p=7} = 15, \tag{12.1}$$

where each factor is the number of residues mod p that avoid both 0 and -2. So there are exactly 15 *admissible twin slots* on the wheel (Fig. 12.1). This is the wheel made to do the work of the notion of an *admissible pattern*: the pair $(0, 2)$ is admissible because, modulo every prime, it misses at least one residue (it never fills all of them, as the inadmissible triple $(0, 2, 4)$ does modulo 3), and the 15 slots are simply where that admissibility lands on the 210-cycle.

The single factor 1 at $p = 3$ deserves a moment, because it is one elementary fact wearing three masks across this book. It equals 1 because the lower member of a twin must avoid both 0 and $-2 \equiv 1$ modulo 3, leaving only $r \equiv 2 \pmod 3$; the upper member $r + 2$ is then $\equiv 1 \pmod 3$, and the multiple of 3 falls between them. So every twin slot reads $2 \to 1$ modulo 3. That same mod-3 split is exactly what the repeating-decimals chapter calls the two cyclic *families* of the wheel when it is written in base 10, and exactly the conductor-3 character $\chi_{3.2}$ of the conductor-lattice chapter. Twin admissibility, the decimal families, and $\chi_{3.2}$ are a single elementary fact about the prime 3, seen from three directions.

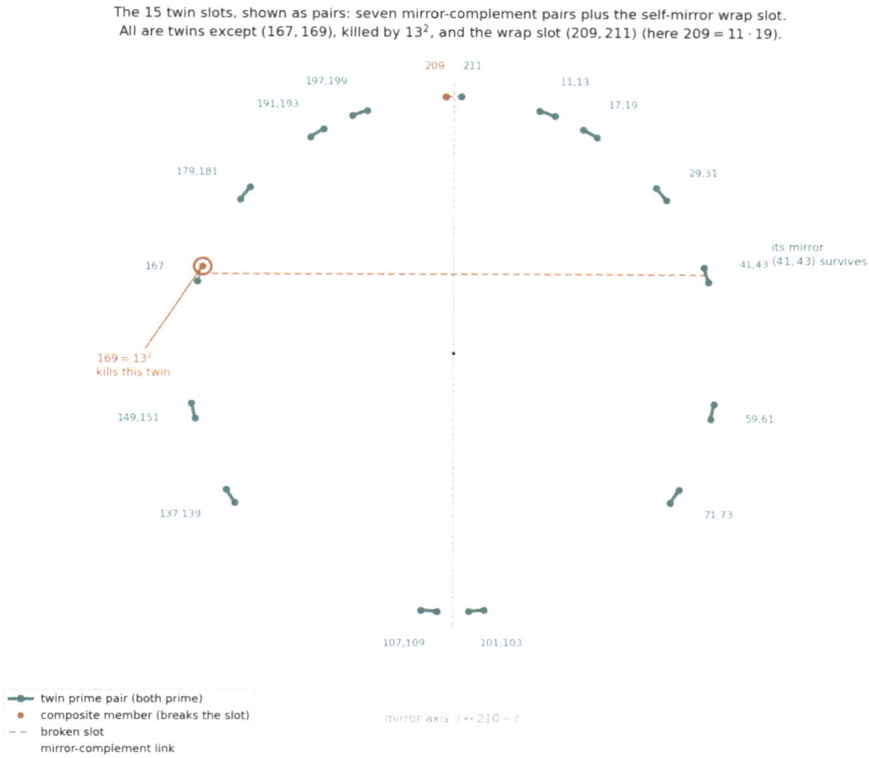

Figure 12.1: The 15 admissible twin slots on the wheel, each drawn as its two members $(r, r+2)$ joined by a bond; faint dotted links connect the mirror-complement pairs across the axis. Each member is colored by what it is—green if prime, red if composite—and a bond is green for a genuine twin, grey for a broken slot. The slots are mirror-symmetric, but the primes are not: $(41, 43)$ (right) is a twin while its mirror $(167, 169)$ (left) is broken at the single composite $169 = 13^2$, with 167 itself still prime. The self-mirror wrap slot at the top is broken at $209 = 11 \cdot 19$, with 211 prime.

12.3 The mirror, broken

The reflection $r \mapsto 210 - r$ of the opening and two-mirrors chapters carries a twin slot to a twin slot: it sends $(r, r+2)$ to $(208 - r, 210 - r)$, again a pair of the right shape. The 15 slots therefore arrange themselves into 7 mirror pairs together with one slot fixed by the reflection—the wrap-around slot $(209, 211)$, sitting on the axis at the top of Fig. 12.1. (This spanning slot is a genuine admissible slot and is counted with the rest: its first representative $(209, 211)$ is broken since $209 = 11 \cdot 19$, but the class holds true twins higher up, the first being $(419, 421)$.) As a pattern of *slots*, the twins are exactly symmetric.

The primes break the symmetry. Looking at the residues themselves, six of the seven mirror pairs hold two genuine twins apiece, but the pair $\{(41, 43), (167, 169)\}$ does not: 41 and 43 are both prime, while across the axis 167 is prime and $169 = 13^2$

is not. The reflection promises a twin on each side; the square 13^2 falls into one slot and quietly removes it, leaving its mirror standing. It is the only interior pair so broken in the first period, and the cause is sharp—13 is the first prime beyond the wheel's own $2, 3, 5, 7$ whose square lands in a twin slot ($11^2 = 121$ misses, having no coprime neighbor two away).

The break is a statement about the small cases, and it is worth being exact about its scope. The *slots* are symmetric; the *asymptotic densities* are symmetric too, since the twins equidistribute among all 15 admissible slots as $x \to \infty$, a mirror slot receiving the same share as its partner. What is *not* symmetric is the finite, concrete arithmetic: which of the literal residues happen to be prime. That is the recurring lesson of the whole book in one picture—the wheel is regular and its symmetries are exact, while the primes scattered on it are not, and the square sitting in $(167, 169)$ is the irregularity made visible.

12.4 The Hardy–Littlewood heuristic

How many twins are there, and how are they shared among the slots? Hardy and Littlewood conjectured in 1923 that the number of twin primes up to x is

$$\pi_2(x) \sim 2C_2 \int_2^x \frac{dt}{(\ln t)^2}, \qquad C_2 = \prod_{p \geq 3} \left(1 - \frac{1}{(p-1)^2}\right) \approx 0.6601618, \tag{12.2}$$

where C_2 is the twin prime constant. The factor in front is the *singular series* for the pair $(0, 2)$: a product of local densities, one per prime, measuring how the admissibility of the pattern translates into a global rate. The wheel is where those local densities live—the 15 surviving slots out of 210 are precisely the $p = 2, 3, 5, 7$ factors of the singular series, made concrete.

Equation (12.2) is only a heuristic, but it is an extraordinarily accurate one. Figure 12.2 compares it with the actual twin count up to 10^7: the prediction tracks the data across four decades, and at $x = 10^7$ there are 58,980 twin pairs against a predicted 58,754, a ratio of 1.004.

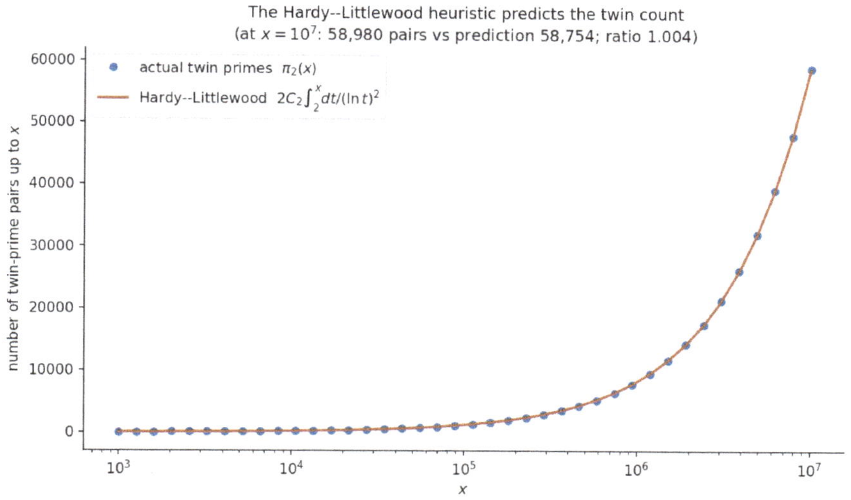

Figure 12.2: The actual count of twin primes $\pi_2(x)$ (points) against the Hardy–Littlewood prediction (12.2) (curve). The heuristic matches the data to within half a percent at 10^7. It predicts the rate with great precision; it does not prove the count is unbounded.

12.5 What this is and is not

The slot structure and the broken mirror are exact and elementary; they are facts. The Hardy–Littlewood formula is a heuristic—a conjecture, not a theorem—and although it agrees with computation to remarkable precision, it is unproven. And the most basic question stands behind both: it is not known whether there are infinitely many twin primes at all. The wheel guarantees 15 admissible slots; it does not guarantee that any single slot is filled infinitely often.

The progress that exists is real but falls short of the target. Zhang proved in 2013 that infinitely many prime pairs differ by at most a fixed bound, and the work of Maynard, Tao, and the Polymath project lowered that bound to 246 [32, 33, 34]. A gap of 2—an actual twin—infinitely often is not among the things proved. So the capstone of this tour ends where the whole subject does: at the primes, and at a wall.

Reproducing the figures

```
import numpy as np
from math import gcd
from sympy import isprime

# the 15 admissible twin slots mod 210, and which are genuine twins
slots = [r for r in range(210)
         if gcd(r,210)==1 and gcd((r+2)%210,210)==1]
print(len(slots))                                    # 15
for r in slots:
```

```
    print(r, r+2, isprime(r) and isprime(r+2))   # (167,169),(209,211) False

# Hardy-Littlewood vs actual twin count up to 1e7
N = 10_000_000
s = np.ones(N+2, bool); s[:2] = False
for i in range(2, int(N**0.5)+1):
    if s[i]: s[i*i::i] = False
twins = np.nonzero(s[:N-1] & s[2:N+1])[0]
from scipy.integrate import quad
C2 = 0.6601618158468695
HL = 2*C2*quad(lambda t: 1/np.log(t)**2, 2, N)[0]
print(len(twins), round(HL), len(twins)/HL)          # 58980  58754  1.0038
```

Part V

Closing

Chapter 13

The Walls

13.1 Why 210?

A natural temptation, having spent a whole book on a single number, is to grant 210 a standing of its own—to feel that $2 \cdot 3 \cdot 5 \cdot 7$ must be doing something the integers themselves care about. It is worth being clear, at the close, that it is not.

The wheel has been a lens, not a law. A law would be a rule the primes obey, something fundamental about them; a lens is only a thing we look through. The number 210 is the second kind of object. The primes do not obey it or notice it. It has been the clearest window we have onto structure that belongs to all the integers, and it earned that role by being small: 210 is the smallest modulus that resolves the first four primes at once, and so the smallest wheel on which the whole early sieve is visible in a single turn—the forty-eight spokes, the mirror across the diameter, the fifteen admissible twin slots, the metronome diagonals of 11, 13, and 17. Every one of these is fixed by the arithmetic of $2 \cdot 3 \cdot 5 \cdot 7$. The lattice is ruled by arithmetic at every point; nothing on it is arbitrary, and that complete determinism is the source of the wheel's clarity. It is also the source of a tempting misreading.

The misreading is to take the tidy patterns of the first turn for laws. They are not laws; they are what the arithmetic of this particular small modulus produces. On the first turn four of the five coprime composites reflect onto primes (only 209 reflects onto the unit 1, not a prime), and the symmetry looks all but total—but that is the arithmetic of 210 speaking about 210, not about the integers, and it does not survive the climb. By the second turn the mirror already pairs two composites with each other, $253 = 11 \cdot 23$ against $377 = 13 \cdot 29$; a primorial or two further up, the clean reflections of the first wheel are gone. Nothing has broken, and nothing was ever arbitrary. The same arithmetic that forces the tidy alignment at 210 forces a different, less tidy one at 2310, and a messier one again at 30030—each fully determined, each particular to its own modulus. What looked like a privilege of 210 was the legibility of a small number, and legibility is not law.

So the structures the wheel made visible—the characters, the explicit formula, the mirror, the singular series, the primon gas—are properties of the integers, reachable from any modulus; 210 only gave the clearest first view of them. And the questions they run into cannot be answered by choosing a better modulus, because they are the same on every wheel. They are the walls of the subject, and the rest of this chapter walks up to them.

13.2 Three walls, one frontier

This book has been, from the first spoke, a work of description. It located the primes on the wheel, measured how evenly they spread, drew the faint bias in their race, found their shadow in the zeros, and counted the slots a twin could occupy. Description, however, is not proof, and three times the description ran all the way up to a fact it could not establish. Those three places are not separate dead ends. They are one frontier seen from three sides.

13.2.1 Whether the primes are forced

The coin-flip model of the opening chapter made the primes look like chance at density $1/\log n$, and indeed they pass almost every test of randomness one can put to them while being completely determined. The explicit formula, in the diffraction and spectrum chapters, showed that the gap between *behaves like* and *is* has an exact address: the zeros of the zeta function. If every nontrivial zero lies on the critical line, the primes' apparent randomness is reined in as tightly as the coin-flip suggests; the Riemann Hypothesis [9] is precisely the assertion that they do. We have checked the first many trillions of zeros and found every one on the line, and we cannot prove the next will be. The spectrum chapter pointed at where a proof might live—a self-adjoint operator whose eigenvalues are the zeros, which would pin them to the line at a stroke—but no such operator is known. The prime race stands on the same wall: its small, stubborn bias is explained, but only by assuming the zeros behave, the generalized form of the same unproven statement.

13.2.2 Whether the twins run out

The wheel guarantees fifteen admissible twin slots and not one more; it guarantees nothing about whether those slots stay filled forever. The Hardy–Littlewood heuristic of the twin-slot chapter predicts that they do—infinitely many twins, at a density the singular series fixes from the very same fifteen-of-210 count [31]—and the prediction tracks the data beautifully. But a heuristic is not a proof, and admissibility is only a necessary condition: the wheel says where a twin *can* sit, never that one *must*. The breakthrough work on bounded gaps showed that *some* gap below a few hundred recurs infinitely often [32, 33]—an extraordinary result, and still a finite distance from the gap of 2 that the slots describe. This is the additive wall: the wheel's multiplicative bookkeeping, for all its precision, cannot by itself reach a statement about pairs.

13.2.3 Whether multiplication can be undone

The two sides of the coin, back in the opening chapter, were never peers. Forming a subprime is a moment's multiplication; recovering the primes it was built from is factoring, for which no efficient method is known. The wheel drew that asymmetry honestly—the duality perfect, the direction not—but it could not prove it permanent. No one has shown that factoring *must* be hard; it is not even known to be as hard as the famous hard problems, and a quantum machine running Shor's algorithm would undo it efficiently in a way no classical method is known to [35]. So even this wall is, strictly, unproven to be a wall. What is certain is only that the easy direction and the hard direction look nothing alike, and that a great deal of the modern world is built on the gap between them holding.

13.2.4 One frontier

Set side by side, the three walls are one thing seen from three angles. The primes are deterministic and yet pseudorandom, and each wall is a place where the precise description the wheel supplies—a density, a set of slots, an easy multiplication—meets a deep fact it cannot prove: that the randomness is genuinely bounded, that the slots are genuinely filled forever, that the multiplication is genuinely irreversible. The Riemann Hypothesis is this frontier on its multiplicative, spectral side; the twin-prime conjecture is the same frontier on its additive side; the hardness of factoring is the same frontier on its algorithmic side. The wheel gave the clearest available view of all three. It cannot pass any of them, and no better-chosen modulus can either.

13.3 Where the wheel leaves us

We set out to see real analytic number theory through a single object, at human scale, and that is what the wheel delivered: the Dirichlet characters as its harmonics, the zeros as its spectrum, the explicit formula as honest interference, the singular series as fifteen surviving slots, the primes as the modes of a gas whose partition function is ζ. None of it was metaphor borrowed to dress up something thinner; it was the standard machinery of the subject, made concrete enough to point at.

The other half of the promise was that the tour would end here—in front of the walls, rather than pretending to pass them. It is worth saying why that is not a disappointment. Knowing exactly where a wall stands, knowing why it stands, and knowing that no clever choice of lens will move it, is not a lesser thing than a proof. It is what honest understanding of an unfinished subject actually looks like. The wheel was a good place to stand. It set us down at the edge of what is known, looking straight at what is not.

Glossary

Wheel The residues modulo $210 = 2 \cdot 3 \cdot 5 \cdot 7$ arranged around a circle; the $\varphi(210) = 48$ residues coprime to 210 are its live *spokes*.

Subprime A composite that survives the wheel — a product of primes all ≥ 11 — as opposed to a genuine prime. The subprimes are the shadow the primes cast through multiplication.

Mirror The reflection $r \mapsto 210 - r$, an involution pairing the 48 spokes into 24 mirror pairs; the elementary shadow of the functional equation $s \mapsto 1 - s$.

Turning Advancing one full period: the spoke r on turn k carries the integer $210k + r$.

Conductor The smallest modulus on which a Dirichlet character is defined; the seven real characters of the wheel are organized by their conductors into a Boolean lattice.

Equidistribution The asymptotic fact that the primes fall in equal share $1/48$ across the 48 spokes (the prime number theorem for arithmetic progressions).

Chebyshev bias The faint, persistent tilt by which some residue classes lead the prime race; governed by the zeros of Dirichlet L-functions.

Explicit formula The exact duality expressing the prime counting function as a sum over the zeros of the zeta function (and, twisted by a character, of L-functions).

Singular series The arithmetic factor in the Hardy–Littlewood conjectures; for twin primes it is the $p = 2, 3, 5, 7$ product made concrete as fifteen admissible slots out of 210.

Admissible slot A residue r for which both r and $r + 2$ are coprime to 210 — a position where a twin-prime pair can sit. There are exactly fifteen.

Primorial A product of the first several primes; 210 is the fourth primorial, $2 \cdot 3 \cdot 5 \cdot 7$.

Primon gas The model in which the primes are the modes of a bosonic gas whose partition function is the zeta function; its fermionic superpartner yields the Möbius function.

The forty-eight units and their mirror partners

The $\varphi(210) = 48$ residues coprime to 210, listed as the 24 mirror pairs $r \leftrightarrow 210 - r$.

$1 \leftrightarrow 209$	$11 \leftrightarrow 199$	$13 \leftrightarrow 197$	$17 \leftrightarrow 193$
$19 \leftrightarrow 191$	$23 \leftrightarrow 187$	$29 \leftrightarrow 181$	$31 \leftrightarrow 179$
$37 \leftrightarrow 173$	$41 \leftrightarrow 169$	$43 \leftrightarrow 167$	$47 \leftrightarrow 163$
$53 \leftrightarrow 157$	$59 \leftrightarrow 151$	$61 \leftrightarrow 149$	$67 \leftrightarrow 143$
$71 \leftrightarrow 139$	$73 \leftrightarrow 137$	$79 \leftrightarrow 131$	$83 \leftrightarrow 127$
$89 \leftrightarrow 121$	$97 \leftrightarrow 113$	$101 \leftrightarrow 109$	$103 \leftrightarrow 107$

Bibliography

[1] G. H. Hardy and E. M. Wright, *An Introduction to the Theory of Numbers*, 6th ed., Oxford University Press, 2008.

[2] H. Davenport, *Multiplicative Number Theory*, 3rd ed., Springer, 2000.

[3] H. Cramér, "On the order of magnitude of the difference between consecutive prime numbers," *Acta Arith.* **2**, 23–46 (1936).

[4] P. G. L. Dirichlet, *Beweis des Satzes, dass jede unbegrenzte arithmetische Progression... unendlich viele Primzahlen enthält*, Abh. Königl. Preuss. Akad. Wiss. (1837), 45–81.

[5] H. L. Montgomery and R. C. Vaughan, *Multiplicative Number Theory I: Classical Theory*, Cambridge University Press, 2007.

[6] M. Rubinstein and P. Sarnak, *Chebyshev's bias*, Experiment. Math. **3** (1994), 173–197. doi:10.1080/10586458.1994.10504289

[7] A. Granville and G. Martin, *Prime number races*, American Mathematical Monthly **113** (2006), 1–33. doi:10.2307/27641834

[8] R. J. Lemke Oliver and K. Soundararajan, *Unexpected biases in the distribution of consecutive primes*, Proc. Natl. Acad. Sci. USA **113** (2016), E4446–E4454. doi:10.1073/pnas.1605366113

[9] B. Riemann, *Über die Anzahl der Primzahlen unter einer gegebenen Grösse*, Monatsber. Berlin Akad. (1859).

[10] H. L. Montgomery, *The pair correlation of zeros of the zeta function*, in *Analytic Number Theory*, Proc. Sympos. Pure Math. **24** (1973), 181–193.

[11] A. P. Guinand, *A summation formula in the theory of prime numbers*, Proc. London Math. Soc. (2) **50** (1948), 107–119.

[12] A. Weil, *Sur les "formules explicites" de la théorie des nombres premiers*, Comm. Sém. Math. Univ. Lund (1952), 252–265.

[13] F. Dyson, *Birds and frogs*, Notices Amer. Math. Soc. **56** (2009), 212–223.

[14] M. V. Berry, *Hearing the music of the primes: auditory complementarity and the siren song of zeta*, J. Phys. A: Math. Theor. **45** (2012), 382001.

[15] B. Mazur and W. Stein, *Prime Numbers and the Riemann Hypothesis*, Cambridge University Press, 2016.

[16] M. R. Watkins, *Fourier analysis and number theory* (expository web resource), `https://empslocal.ex.ac.uk/people/staff/mrwatkin/zeta/NTfourier.htm`.

[17] L. Csoka, *The Fourier transform of the non-trivial zeros of the zeta function*, arXiv:1712.08434 (2017).

[18] A. M. Odlyzko, *On the distribution of spacings between zeros of the zeta function*, Math. Comp. **48** (1987), 273–308. doi:10.1090/S0025-5718-1987-0866115-0

[19] Z. Rudnick and P. Sarnak, *Zeros of principal L-functions and random matrix theory*, Duke Math. J. **81** (1996), 269–322. doi:10.1215/S0012-7094-96-08115-6

[20] O. Bohigas, M. J. Giannoni, and C. Schmit, *Characterization of chaotic quantum spectra and universality of level fluctuation laws*, Phys. Rev. Lett. **52** (1984), 1–4. doi:10.1103/PhysRevLett.52.1

[21] M. V. Berry and J. P. Keating, *The Riemann zeros and eigenvalue asymptotics*, SIAM Rev. **41** (1999), 236–266. doi:10.1137/S0036144598347497

[22] A. Connes, *Trace formula in noncommutative geometry and the zeros of the Riemann zeta function*, Selecta Math. (N.S.) **5** (1999), 29–106. doi:10.1007/s000290050042

[23] B. L. Julia, *Statistical theory of numbers*, in *Number Theory and Physics*, eds. J. M. Luck, P. Moussa, M. Waldschmidt, Springer Proc. Phys. **47** (1990), 276–293.

[24] D. Spector, *Supersymmetry and the Möbius inversion function*, Comm. Math. Phys. **127** (1990), 239–252.

[25] I. Bakas and M. J. Bowick, *Curiosities of arithmetic gases*, J. Math. Phys. **32** (1991), 1881–1884. doi:10.1063/1.529511

[26] S. A. Hartnoll and M. Yang, *The conformal primon gas at the end of time*, J. High Energy Phys. **07** (2025), 281 (generalizing the primon gas: a Dirichlet L-function as a charged-primon partition function). doi:10.1007/JHEP07(2025)281

[27] D. Schumayer and D. A. W. Hutchinson, *Physics of the Riemann hypothesis*, Rev. Mod. Phys. **83** (2011), 307–330. doi:10.1103/RevModPhys.83.307

[28] F. Mertens, *Ein Beitrag zur analytischen Zahlentheorie*, J. Reine Angew. Math. **78** (1874), 46–62.

[29] G. Tenenbaum, *Introduction to Analytic and Probabilistic Number Theory*, 3rd ed., AMS, 2015.

[30] C. F. Gauss, *Disquisitiones Arithmeticae* (Leipzig, 1801), Section VI —the periods of decimal fractions and their dependence on the order of 10 modulo the denominator. English translation by A. A. Clarke, Yale University Press, 1966.

[31] G. H. Hardy and J. E. Littlewood, *Some problems of 'Partitio numerorum' III: On the expression of a number as a sum of primes*, Acta Math. **44** (1923), 1–70. doi:10.1007/BF02403921

[32] Y. Zhang, *Bounded gaps between primes*, Ann. of Math. **179** (2014), 1121–1174. doi:10.4007/annals.2014.179.3.7

[33] J. Maynard, *Small gaps between primes*, Ann. of Math. **181** (2015), 383–413. doi:10.4007/annals.2015.181.1.7

[34] D. H. J. Polymath, *Variants of the Selberg sieve, and bounded intervals containing many primes*, Res. Math. Sci. **1**:12 (2014). doi:10.1186/s40687-014-0012-7

[35] P. W. Shor, *Polynomial-time algorithms for prime factorization and discrete logarithms on a quantum computer*, SIAM J. Comput. **26**, 1484–1509 (1997). doi:10.1137/S0097539795293172

www.ingramcontent.com/pod-product-compliance
Lightning Source LLC
LaVergne TN
LVHW070856210926
848482LV00016B/949